Development of Rural Women Entrepreneurship

The Author

Dr. (Mrs) Gyanmudra started her career as a Psychologist (Scientist D) Defence Research Development Organisation in the Ministry of Defence, Government of India, and worked for one and a half decade. Currently she is working as Head, Centre for Human Resource Development in National Institute of Rural Development, Hyderabad, Government of India.

Development of Rural Women Entrepreneurship
An Analysis of Social and Psychological Dimensions

Dr. Gyanmudra
Project Coordinator & Head
Centre for Human Resource Development
National Institute of Rural Development
Hyderabad

2013
Daya Publishing House®
A Division of
Astral International Pvt. Ltd.
New Delhi – 110 002

© 2013, AUTHOR
ISBN 9789351241140

Published by : **Daya Publishing House®**
A Division of
Astral International Pvt. Ltd.
– ISO 9001:2008 Certified Company –
4760-61/23, Ansari Road, Darya Ganj
New Delhi-110 002
Ph. 011-43549197, 23278134
E-mail: info@astralint.com
Website: www.astralint.com

Laser Typesetting : **Classic Computer Services**
Delhi - 110 035

Printed at : **Salasar Imaging Systems**
Delhi - 110 035

PRINTED IN INDIA

Acknowledgements

First and foremost I would like to extend my gratitude to Mr. Mathew C. Kunnumkal, Ex.Director General, NIRD for giving me the opportunity to undertake the study on Identifying Facilitating and Hindering Factors of Women Entrepreneurs: An Analysis of Social and Psychological Dimensions.

I would like to thank, Dr. M. V. Rao, Director General, NIRD for his constant support and encouragement.

I would also like to thank all the associates who helped at various stages of the study from preparation and testing of schedules to data collection to analysis and report writing. Thank to Anju Helen Bara, Mr. Sampath Reddy and K. Papamma for editing and proof reading. Many thanks to the field investigators who helped me in collecting data and translating the language during field work.

Last but not the least, I would like to thank various State level officials in Tamil Nadu, field level officials and other members who gave us their time during various interactions and helped us to capture the insights of their life.

A special thanks to my husband, Mr. Somesh Kumar, IAS for his constant encouragement which played a significant role in the completion of this work.

I would also like to thank Publication House for printing the book. Finally, I am grateful to all my colleagues for supporting and encouraging me in this endeavour.

Dr. Gyanmudra

Foreword

I am pleased to introduce this book *'Development of Rural Women Entrepreneurship: An Analysis of Social and Psychological Dimensions'* by Dr Gyanmudra, currently Head, Centre for Human Resource Development, National Institute of Rural Development, Hyderabad. Developing countries like India need to accord priority to entrepreneurship and skill development to bring in the desired levels of development. Rural areas in particular are in need of entrepreneurship opportunities. It has been observed that women, who have the responsibility of family and children, need entrepreneurship skills to improve their living conditions and to give better future to their children. This study is an important contribution in understanding the condition of women entrepreneurship in villages of India.

This study examines the facilitating and inhibiting factors in the growth of women entrepreneurship and also looks at various interventions of Government and non-government agencies which help women to achieve higher incomes and economic freedom. It also brings out interesting social and psychological aspects of entrepreneurship development process in our rural areas. The case studies presented in this study show factors which aid the process of women entrepreneurship.

This study is an important addition to the existing literature on women entrepreneurship development in India.

Dr. M.V.Rao

Director General

National Institute of Rural Development

Hyderabad

Contents

List of Figures

List of Tables

1

Introduction

Entrepreneurship has been an indispensable factor contributing for the development of many countries. It is the dearth of entrepreneurship, which has been the foremost factor for backwardness of developing countries like India. Today's world is changing at startling pace. Political and economic transformations seem to be occurring everywhere – as countries convert from command to demand economics, dictatorships move toward democracy, and monarchies build new civil institutions. These changes have created economic opportunities for women who want to own and operate business. Today, women in advanced market economics own more than 25 per cent of all business. In some regions of the world, transformation to a market economy threatens to sharpen gender inequality. Some of these changes are simply the legacy of a gender imbalance that existed prior to political and economic reform. Other changes reflect a return to traditional norms and values that relegated women to a secondary status. As countries become more democratic, gender inequalities lesson: thus, offering a more productive atmosphere for both sexes.

To choose the definition of entrepreneurship most appropriate in the context of rural area context, it is important to bear in mind the entrepreneurial skills that will be needed to improve the quality of life for individuals, families and communities and to sustain a healthy economy and environment. Taking this into consideration, we will find that each of the traditional definitions has its own weakness (Tyson, Petrin, & Rogers, 1994). The first definition leaves little room for innovations that are not on the technological or organizational cutting edge, such as, adaptation of older technologies to a developing-country context, or entering into export markets already tapped by other firms. Defining entrepreneurship as risk-taking neglects other major elements of what we usually think of as entrepreneurship, such as a well-developed ability to recognize unexploited market opportunities. Entrepreneurship as a stabilizing force limits entrepreneurship to reading markets disequilibria, while entrepreneurship defined as owning and operating a business, denied the possibility of entrepreneurial behaviour by non-owners, employees and managers who has no equity stake in the business.

Therefore, the most appropriate definition of entrepreneurship that would fit into the rural development context, argued here, is the broader one, the one which defines entrepreneurship as: 'a force that mobilizes other resources to meet unmet market demand', 'the ability to create and build something from practically nothing', 'the process of creating value by pulling together a unique package of resources to exploit an opportunity'. It combines definitions of entrepreneurship by Jones and Sakong, 1980; Timmons, 1989; Stevenson, *et al.*, 1985.

Rural developments are more than ever before linked to entrepreneurship. Institutions and individuals promoting rural development now see entrepreneurship as a strategic development intervention that could accelerate the rural development process. Furthermore, institutions and individuals seem to agree on the urgent need to promote rural enterprises: development agencies see rural entrepreneurship as an enormous employment potential; politicians see it as the key strategy to prevent rural unrest; farmers see it as an instrument for improving farm earnings; and women see it as an employment possibility which provides autonomy, independence and a reduced need for social support. To all these groups, however, entrepreneurship stands as a vehicle to improve the quality of life of individuals, families and communities and to sustain a healthy economy and environment.

The research which tries to explain, by personal traits and/or other social aspects, why certain individuals become entrepreneurs, has not yet produced convincing results. Consequently, a widely accepted view is the following: while personal characteristics as well as social aspects clearly play some role, entrepreneurship and entrepreneurs can also be developed through conscious action. Development of entrepreneurs and of entrepreneurship can be stimulated through a set of supporting institutions and through deliberate innovative action which stimulates changes and fully supports capable individuals or groups. It is argued, that controllable variables such as a stable system of property rights and freedom of action in the economic sphere, availability of other inputs in the economy (besides entrepreneurship) as well as education and training, contribute significantly to the development of entrepreneurship. Therefore, policies and programmes designed specifically for entrepreneurship promotion can greatly affect the supply of entrepreneurs and thus indirectly represent an important source of entrepreneurship.

Given the background, women entrepreneurs face several barriers when entering into business and recognizing challenges: 1) Behavioural barriers 2) Personal barriers 3) Socio-cultural barriers 4) Educational barriers 5) Economic and Political barriers 6) Financial barriers 7) Personnel or Managerial barriers and 8) Marketing barriers.

It is difficult to theorize, in general, terms, the phenomenon of entrepreneurship. Almost each of the entrepreneurs is a case by itself, and rarely are they identical in any respect. The better methodology for studying the evolution and development of entrepreneurship would be the "case-study" analyses which could give a penetrating insight into the cases from their own perspectives. It is in this background the present study seeks to make a case study of women entrepreneurs located in Tamil Nadu.

Women entrepreneurs engage in varied production/ trade activities including Gem cutting, Grocery, Medicine Wholesale, Paper Bag, Tailoring, Sottu Neelam, Flour Mill, Book Binding, Bakery, and Vermi Compost Unit will be analysed.

Women entrepreneurs are divided into three categories:

1. The first group consists of women with educational and professional qualification. Group takes the initiative and manages the business. Women entrepreneurs who have the basic managerial training and educational qualification go for the medium and large scale units.

2. The second group consists of those women entrepreneurs who do not have education or any formal training in management but have developed practical skills required for the small scale sector choose that product with which they are familiar. For example garments, dolls, handicraft items etc.

3. The third group of women entrepreneurs works in cities and slums to help women with lower means of livelihood. They need Government support in marketing as well as in getting finance at concessional rates.

Women are entering into entrepreneurship even while facing socio-cultural, economic, technical, financial and managerial difficulties. Women entrepreneurship movement can gain momentum by providing encouragement, appropriate awareness, training, environment and support. This would definitely enhance socio-economic status of women.

In most of the developing countries today, more and more emphasis is laid on the need for development of women and their active participation in the main stream of development process. It is also widely recognized that apart from managing household, bearing children, rural women bring income with productive activities ranging from traditional work in the fields to working in factories or running small and petty businesses. They have also proven that they can be better entrepreneurs and development managers in any kind of human development activities. Therefore, it is important and utmost necessary to make rural women empowered in taking decisions to enable them to be in the central part of any human development process. The empowerment of women also considered as an active process enabling women to realize their full identity and power in all spheres of life.

Women's empowerment is a global issue and discussions on women's rights are at the forefront of many formal and informal campaigns worldwide. This is because women have been regarded as the nuclei of a nation and the builders and moulders of its destiny. The position and status of women in any society is an index of its civilization. The concept of empowerment received momentum after the Bangladesh experiments of the 1980's led by Prof. Muhammed Yunus, founder of the Grameen Bank, got worldwide acclamation. Following the example, various countries including India introduced empowerment models as a strategy for women empowerment. As a result, various women-led organizations were formed with the common focus of poverty reduction and empowerment of women.

The empowerment of women through entrepreneurship would reap benefits not only for the individual women but also for the family and community as a whole. Empowering women is not just for meeting their economic needs but also more holistic social development. Rural women are more vulnerable in comparison to urban women, because the urban women have wide scope of activities around them to explore but rural women do not get such opportunities to make use of their economic potential. Entrepreneurial skill and marketing talent are to be given to these women, through proper training programmes.

The higher economic status, self-reliance and self-esteem imbue them with power to make changes and choices about their lives. The choices now made extend to education, housing, health-care, and political participation. Thus as a recent on-going study clearly states "in order for a woman to be empowered, she needs access to the material, human, and social resources necessary to make strategic choices in her life." By many women being able to make their own choices they become agents of change who, in turn, are able not only to challenge, but also to organize other women to challenge the social, economic, religious, and political structures of injustice that keep them down. The empowerment that is provided by financial access creates further synergies that lead to the acquisition of education and literacy; business training and management; and access to information.

The gap in research on this topic is striking since studies on entrepreneurship in general attribute great importance to psychological factors and the role of the family and society in shaping individual motivation and behaviour. It is found that through the motivational route, personality influences entrepreneurial behaviour (Singh, 1997). Ingredients like need for achievement, economic independence, the autonomy are essential elements for the success of an entrepreneur (Pear, 1989). On the psychological side, willingness to take risks, ambition, a strong desire for individual achievement, and persistence are considered some of the main traits (Kaza, 1996). Especially with those who are unaccustomed to taking risks, the fear of failure (psychological) and of peer opinion (social) are predominant at the entry level. The key barrier that a woman entrepreneur has to overcome is the fear of risk (Histrich, 1986). In short, an entrepreneur is very different from a non-entrepreneur in social and psychological disposition (Rank, 1996). Socio-psychological factors which have significant association with entrepreneurial success has not been studied much. Therefore the present study has been taken up with the following objectives.

1.2 Objectives

- ☆ To identify the psycho-social factors that governs enterprise development
- ☆ To identify facilitating and hindering factors of Women Entrepreneurs
- ☆ To collect information on specific actions and support measures promoting female entrepreneurship in the State
- ☆ To identify good practices in the promotion of female entrepreneurship.

2

Development of Rural Women Entrepreneurship

Women in India constitute around 50 per cent of the total population and comprise one-third of the labour force. It is, therefore, important that, when considering the economic development of this segment of the population, due attention is given to their socioeconomic empowerment. India's first Prime Minister, Pandit Jawaharlal Lal Nehru, stated," In order to awaken the people, it is the woman who has to be awakened. Once she is on the move, the household moves, the village moves, the country moves, and thus, we build the India of tomorrow." While addressing developmental issues, recognition that both men and women are equal partners of any development is necessary with both having equal opportunities to share the benefits (Anon, 1997).

The society of women is the foundation of good manners; of course a pre-requisite to achieve brilliant results especially for success in business. The increasing trend developed among the women to be self-employed suggests that time is not far away when women factor would also have an important role in the economic growth of the country. Possessing the natural gift of politeness, women entrepreneurs, and if provided the level ground, are expected to bring new milestones to this country.

The entrepreneurial orientation to rural development accepts entrepreneurship as the central force of economic growth and development, without it other factors of development will be wasted or frittered away. However, the acceptance of entrepreneurship as a central development force by itself will not lead to rural development and the advancement of rural enterprises. What is needed in addition is an environment enabling entrepreneurship in rural areas. The existence of such an environment largely depends on policies promoting rural entrepreneurship. The effectiveness of such policies in turn depends on a conceptual framework about entrepreneurship, *i.e.*, what it is and where it comes from.

2.1 Entrepreneurship Concept

The entrepreneurship concept, what it means and where it comes from, is the foundation for policies promoting entrepreneurship and the key to understanding the role of entrepreneurship in development.

Defining entrepreneurship is not an easy task. To some, entrepreneurship means primarily innovation, to others it means risk-taking to others a market stabilizing force and to others it means starting, owning and managing a small business. Accordingly, the entrepreneur is then viewed as a person who either creates new combinations of production factors such as new methods of production, new products, new markets, finds new sources of supply and new organizational forms; or as a person who is willing to take risks; or a person who, by exploiting market opportunities, eliminates disequilibrium between aggregate supply and aggregate demand, or as one who owns and operates a business (Tyson, Petrin, Rogers, 1994).

To choose the definition of entrepreneurship most appropriate in the context of rural area context, it is important to bear in mind the entrepreneurial skills that will be needed to improve the quality of life for individuals, families and communities and to sustain a healthy economy and environment. Taking this into consideration, we will find that each of the traditional definitions has its own weakness (Tyson, Petrin, Rogers, 1994). The first definition leaves little room for innovations that are not on the technological or organizational cutting edge, such as, adaptation of older technologies to a developing-country context, or entering into export markets already tapped by other firms. Defining entrepreneurship as risk-taking neglects other major elements of what we usually think of as entrepreneurship, such as a well-developed ability to recognize unexploited market opportunities. Entrepreneurship as a stabilizing force limits entrepreneurship to reading markets disequilibria, while entrepreneurship defined as owning and operating a business, denies the possibility of entrepreneurial behaviour by non-owners, employees and managers who have no equity stake in the business. Therefore, the most appropriate definition of entrepreneurship that would fit into the rural development context, argued here, is the broader one, the one which defines entrepreneurship as: "a force that mobilizes other resources to meet unmet market demand", "the ability to create and build something from practically nothing", "the process of creating value by pulling together a unique package of resources to exploit an opportunity".

Entrepreneurship so defined, pertains to any new organization of productive factors and not exclusively to innovations that are on the technological or organizational cutting edge, it pertains to entrepreneurial activities both within and outside the organization. Entrepreneurship need not involve anything new from a global or even national perspective, but rather the adoption of new forms of business organizations, new technologies and new enterprises producing goods not previously available at a location. This is why entrepreneurship is considered to be a prime mover in development and why nations, regions and communities that actively promote entrepreneurship development, demonstrate much higher growth rates and consequently higher levels of development than nations, regions and communities whose institutions, politics and culture hinder entrepreneurship.

An entrepreneurial economy, whether on the national, regional or community level, differs significantly from a non-entrepreneurial economy in many respects, not only by its economic structure and its economic vigorousness, but also by the social vitality and quality of life which it offers with a consequent attractiveness to people. Economic structure is very dynamic and extremely competitive due to the rapid creation of new firms and the exit of 'old' stagnant and declining firms. It is populated with rapidly growing firms, gazelles as they are called in the literature of entrepreneurship. Gazelles are the key to economic development. Dynamic entrepreneurs look for growth; they have not only a vision but also capability of making it happen. They think and act globally, look for expansion, rely on external resources, seek professional advice or they work with professional teams. They challenge competitors instead of avoiding them and take and share risks in a way that leads to success. In this way economic vitality of a country largely depends on the overall level of entrepreneurial capacity, *i.e.*, on its ability to create rapidly growing companies, gazelles. Equally important is the speed by which new small companies are created. These phenomena explain why countries, regions and communities with a similar number of large and small firms show a completely different economic vitality.

Economic vitality of a country is no doubt a necessary condition for social vitality. Without it other important factors that make living attractive in certain areas, such as education, health, social services, housing, transport facilities, flow of information and so on, cannot be developed and sustained in the area in the long run.

Entrepreneurial orientation to rural development, contrary to development based on bringing in human capital and investment from outside, is based on stimulating local entrepreneurial talent and subsequent growth of indigenous companies. This in turn would create jobs and add economic value to a region and community and at the same time keep scarce resources within the community. To accelerate economic development in rural areas, it is necessary to increase the supply of entrepreneurs, thus building up the critical mass of first generation entrepreneurs, who will take risks and engage in the uncertainties of a new venture creation, create something from practically nothing and create values by pulling together a unique package of resources to exploit an opportunity. By their example they will stimulate an autonomous entrepreneurial process, as well as a dynamic entrepreneurship, thereby ensuring continuous rural development.

It is important to stress that rural entrepreneurship in its substance does not differ from entrepreneurship in urban areas. Entrepreneurship in rural areas is finding a unique blend of resources, either inside or outside of agriculture. This can be achieved by widening the base of a farm business to include all the non-agricultural uses that available resources can be put to or through any major changes in land use or level of production other than those related solely to agriculture. Thus, a rural entrepreneur is someone who is prepared to stay in the rural area and contribute to the creation of local wealth. To some degree, however, the economic goals of an entrepreneur and the social goals of rural development are more strongly interlinked than in urban areas. For this reason entrepreneurship in rural areas is usually community based, has strong extended family linkages and a relatively large impact on a rural community.

2.2 Sources of Entrepreneurship

From the policy viewpoint the promotion of entrepreneurship, the understanding where entrepreneurship comes from, is as equally important as understanding the concept of entrepreneurship. It indicates where the governments, national, regional or local, should target their promotional efforts. If entrepreneurial skills, for example, are innate, active promotion policies have a small role to play. If instead, certain entrepreneurial characteristics are trained, then active promotion policies can contribute to entrepreneurship development in the community in the region and in the nation, since entrepreneurial skills can be acquired through training.

2.3 Rural Entrepreneurship

Many examples of successful rural entrepreneurship can already be found in literature. Diversification into non-agricultural uses of available resources such as tourism, blacksmith, carpentry, spinning, etc. as well as diversification into activities other than those solely related to agricultural usage, for example, the use of resources other than land such as water, woodlands, buildings, available skills and local features, all fit into rural entrepreneurship. The entrepreneurial combinations of these resources are, for example: tourism, sport and recreation facilities, professional and technical training, retailing and wholesaling, industrial applications (engineering, crafts), servicing (consultancy), value added products from meat, milk, wood, etc. and the possibility of off-farm work. Equally entrepreneurial are new uses of land that enable a reduction in the intensity of agricultural production, for example, organic production.

Dynamic rural entrepreneurs can also be found. They are expanding their activities and markets and they find new markets for their products and services beyond the local boundaries.

The evidence shows that there are many activities in rural areas pursued by female entrepreneurs such as: trade, food processing, handicrafts, production of basic consumer articles, catering, running tourist establishments, and bed and breakfast arrangements. However, compared to male entrepreneurs, female entrepreneurs in rural areas still tend to be limited to what have traditionally been viewed as women's activities. Also the scale of their entrepreneurial operation tends to be smaller when compared with male entrepreneurs.

Although agriculture today still provides income to rural communities, rural development is increasingly linked to enterprise development. Since national economies are more and more globalized and competition is intensifying at an unprecedented pace, affecting not only industry but any economic activity including agriculture, it is not surprising that rural entrepreneurship is gaining in its importance as a force of economic change that must take place if many rural communities are to survive. However, entrepreneurship demands an enabling environment in order to flourish.

2.4 Environment Conducive to Entrepreneurship

Behind each of the success stories of rural entrepreneurship there is usually some sort of institutional support. Besides individual or group entrepreneurial initiative the enabling environment supporting these initiatives is of utmost importance.

The creation of such an environment has already been started at the national level with the foundation policies for macro-economic stability and for well-defined property rights as well as international orientation. Protection of the domestic economy hinders instead of fostering entrepreneurship. National agricultural policies such as price subsidies to guarantee minimum farm incomes and the keeping of land in production during over-productive periods are definitely counter-productive to entrepreneurship. The long run solution for sustainable agricultural development is only one, *i.e.'* competitive agriculture. While prices can set the direction, entrepreneurs who will meet the challenge of increasingly demanding international markets and who will find profitable alternative uses of land, alternative business opportunities and so on are needed. Therefore, policies and programmes targeted more specifically at the development and channeling of entrepreneurial talent, are needed. Policies for boosting the supply of entrepreneurs, policies for developing the market for other inputs into successful entrepreneurship, policies for increasing the effectiveness of entrepreneurs and policies for multiplying demand for entrepreneurship can significantly speed up entrepreneurial activities at national, regional and community levels.

Other institutions that can make a difference to rural development based on entrepreneurship are agricultural extension services. However, to be able to act in this direction, they too must be entrepreneurial minded. They must see agricultural activities as one of many possible activities that contribute to rural development. They must seek new entrepreneurial uses of land and support local initiative in this respect. While tradition is important it is nevertheless dangerous to be over-occupied with the past, otherwise the rural community may turn into a nostalgia-driven society. Networking between different agencies involved in the promotion of rural development through entrepreneurship, by pooling together different sources and skills, by reaching a greater number of future entrepreneurs and by assisting a greater number of local entrepreneurial initiatives, can have a much more positive effect on rural development than when each agency is working on its own.

To summarize, policy implications for rural entrepreneurship development are:

☆ Sound national economic policy with respect to agriculture, including recognition of the vital contribution of entrepreneurship to rural economic development

☆ Policies and special programmes for the development and channeling of entrepreneurial talent

☆ Entrepreneurial thinking about rural development, not only by farmers but also by everyone and every rural development organization

☆ Institutions supporting the development of rural entrepreneurship as well as strategic development alliances

2.5 Women Entrepreneurs

Women entrepreneurs, as research demonstrates, may do things differently. For example, in comparison to male entrepreneurs, women tend to work more in teams, are less self-centred and personal ego to them is less important than success of the organization or business idea they are pursuing.

However, there is no difference in characteristics such as achievement, autonomy, aggression, independence and benevolence between female and male entrepreneurs (Histrich and Brush, 1986). Also, no differences were found in risk taking propensity of male and female entrepreneurs. However, we do need to talk explicitly about women entrepreneurs. It should be stressed that rural women can encounter many constraints when trying to take part in the transformation process. Rural areas tend to be more traditional in regard to the gender issue. In rural areas, the gender issue is usually a much stronger hindering factor to potential female entrepreneurs than it is in urban areas, their self-esteem and managerial skills being lower when compared to urban women and access to external financial resources more difficult than in urban areas. Therefore, special programmes of assistance (technical and financial) to overcome these constraints should be developed and designed to meet the needs of rural women in order to be able to take an active part in entrepreneurial restructuring of their communities, to start to develop their own ventures, to expand their already existing businesses, or to function as social entrepreneurs since their number today is still below the potential one.

This belief is the one for which we as trainers are responsible to bring to rural women in addition to trying to put in place all factors crucial for rural women to enter into entrepreneurial activities. Without it, entrepreneurial opportunities will not be seen, they will be lost and then the role of women in rural development will be much below their potential.

Developing entrepreneurial culture entails the development of a positive attitude towards new and unforeseen circumstances. However, shifting the attitude towards change does not suffice for rural entrepreneurship to thrive. What is important is seeing and seizing entrepreneurial opportunity. Too often, potential entrepreneurs are hindered by the lack of motivation to take risks, the non-availability of information on what is a good business opportunity and the lack of skills/knowledge on how to set up a new enterprise. Even if these constraints were alleviated, it is still not certain that rural areas would embark on the sustainable path of development due to what is involved in new firm creation. What is needed is the development of institutions supporting entrepreneurial restructuring in rural areas.

2.6 Development of a New Venture and Growth

The operating and managing of a business demand certain skills in finance, organization, production, marketing and distribution of products. The entrepreneurs should acquire these skills and should continuously up-date them in order to ensure/maintain their competitive position based on a well-defined long-term business strategy. Too often small firms suffer from managing their businesses on a day to day basis and not realizing their potential for growth. Training inputs therefore concentrate on all aspects of managing the business and on the strategies for future growth:

☆ Managerial skills

☆ Recording, measuring and controlling business performance

☆ Strategies for future growth

2.7 Institution Building

For entrepreneurship to flourish, a supportive environment is needed. Therefore, those who are involved in fostering rural entrepreneurship should not only influence national/regional development policies, but also facilitate the development of national and regional supporting institutions. Training inputs focus on:

☆ The role of national and local governments and non-governmental institutions in promoting entrepreneurial restructuring of rural areas

☆ Financial institutions: commercial banks, special funds (Consortia) and venture capital companies

☆ Networking: understanding the concept, goal, purpose, benefits of networking and developing a strategy to set up the network

☆ Economic development alliances between potential entrepreneurs, new firm formation, community/regional infrastructure and the existing private sector

2.8 The Role of Trainers/Facilitators in the Entrepreneurial Development Process

The above-mentioned four inter-related training inputs embrace the total process whereby the entrepreneurial and managerial capabilities of would-be- entrepreneurs and an environment conducive to entrepreneurship, are developed. Trainers facilitating the implementation of the training programme on entrepreneurship development as outlined here should therefore possess certain behavioural and managerial skills in order to initiate, organize and monitor this process effectively.

Training thus focuses on:

☆ The role of the trainer/facilitator

☆ Qualifications of the trainer/facilitator

☆ Skills of the trainer/facilitator

☆ Training of the trainer/facilitator

The best training results are achieved when participants first gain an understanding of the total process involved in entrepreneurship development and afterwards receive in-depth training on each of the four inter-related processes outlined above.

3

Methodology

The study is based on original data and information generated by using different methods of data collection. This study is to investigate the factors that facilitate as well as hinder the women entrepreneurship. The study is also aimed at the intervention of the both government and non-government agencies that have driven women into financial sustainability and the women's role in integrated community development. All enterprises in the sampling areas chosen by stratified random sampling technique. The survey involves listing of all enterprises through SGSY scheme and interviewing of a sample of 90 entrepreneurs. At the District level, using Questionnaire/personal interview and in-depth case study method from a representative sample of women was conducted. Detailed information was gathered through the case studies on the beginning and development of the enterprise, the problems the women faced at different stages of growth, the prospects for further development and, diversification and the external environment in which they are working.

3.1 Study Area

After the consultation of State Rural Development Officials the two well performing districts were selected to understand the resources and difficulties faced by the entrepreneurs. The present study was carried out in the districts of Coimbatore and Tiruchirapalli in the State of Tamil Nadu. The nine blocks covered in Coimbatore were Pollachi, Kinathukadavu, Anaimalai, Karamadai, P.N. Palayam, Udamalpet, Gudimangalam, Palladam, Tirupur. From Tiruchirapalli district, eleven Blocks, *i.e.* Manikandam, Thiruverumbur, Karumandapam, Andhanallur, Musiri, Lalgudi, Thottiyam, Marungapuri, Uppliyapuram, Edipattipudur, Thurayur were selected for primary data collection. From Coimbatore district, 44 individual women entrepreneurs and from Tiruchirapalli district, 46 individual cases were analysed in detail. The total sample is 90.

There was a discussion with the employees of different departments, *i.e.*, Social welfare department and Women and child welfare department and others, in order to get the supportive data. There was an in depth interview with Government officials at various levels to get qualitative insights.

3.2 District profile of Tiruchirappalli

Tiruchirappalli district is a centrally located district in Tamil Nadu State, bounded by Perambalur District in North, Pudukkottai and Dindikul Districts in South, Thanjavur district in East and Karur district in West. The total area of the district is 4,403Sq.km. The total population of the district is 24,09,466 which accounts for 12,08,634 of the male population and 12,00,832 of the female population. According to the census 2001 the total rural population was 15,77,204 and the urban population was 11,36,162.

About 446 major trade industries, 1628 minor trade industries and 798 other trade industries are there. The other important industries are textile, cement, sugar etc and there is considerable number of cotton, woolen, silk and polyester hanks production in the district. The principal agricultural crops are rice, millets, cereals, pulses, sugarcane, groundnut, gingely. The area is mainly irrigated by government canals and wells. In some parts of the district private canals and tube wells are also in use. There are 8 veterinary hospitals and 44 veterinary dispensaries in the district. The livestock like sheep, goat and horses etc; are there. One dairy and 215 milk co-operative societies are there for which the SHG members sell the milk.

There are 1800 women welfare co-operative societies, 2 training centers and 7029 women development groups. Local bodies like corporations, municipalities, panchayat unions, town panchayats and village panchayats exist in the district. All the villages are fully electrified. There are 14 blocks in the district from which 11 blocks have been selected for the present study.

3.2.1 Wage-Employment Generated

Wage-employment generated during 2006-2007 through Wage Employment Schemes of Ministry of Rural Development and other Development Schemes of Central Government:

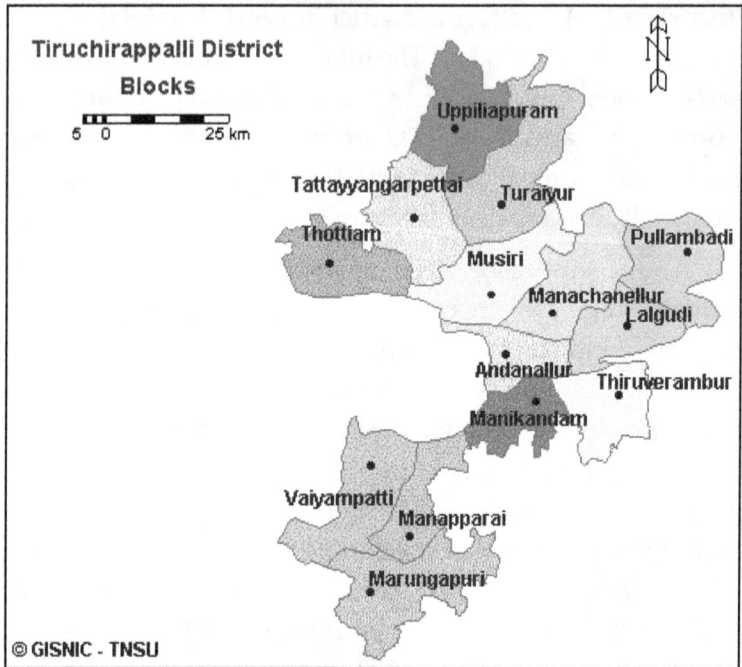

Figure 3.1: Map of Tiruchirappalli district blocks.

3.2.2 Self-Employment Generated

Self-employment generated during 2006-2007 through various programmes of Ministry of Rural Development and other Schemes of Central Government, States and N.G.O.s etc.

3.3 District Profile of Coimbatore

Coimbatore is one of the largest districts in the state of Tamil Nadu. The district is divided into 3 revenue divisions. Coimbatore, Pollachi and Tirupur and 9 taluks comprising of 482 revenue villages. The total population of Coimbatore district is 42,71,856 with 21,76,031 males and 14, 15,653 females. According to the census 2001 the rural population was 14,15,653 and the urban population was 28,20,203.

Table 3.1: Schemes of Central Government (Wage-employment).

| Sl.No. | Name of Scheme and Organisation | Rural/ Urban Areas | Expenditure (Rs. Lakh) | | Person Days of Employment | | |
			Wage Expenditure	All/ Total Expenditure	Male	Female	Total
1	2	3	4	5	5	7	8
1.	S.G.R.Y.	Rural	702.635	1171.058	2439704	12198521	14638225

Source: Ministry of Rural Development, Government of India, http://rural.nic.in/

Table 3.2: Schemes of Central Government (Self-employment).

| Sl.No. | Name of Scheme and Ministries/State/ N.G.O.s | Rural/ Urban Areas | Expenditure (Rs. Lakh) | | Person Days of Employment | | |
			Wage Expenditure	All/ Total Expenditure	Male	Female	Total
1.	S.G.S.Y. (Ministry of Rural Development)	Rural–E.A. to 115 S.H.G.s	97.300	97.300	–	–	–

Source: Ministry of Rural Development, Government of India, http://rural.nic.in/

The people working in the urban areas are greater in number than the people working in rural areas. There are also cultivators, agricultural labourers and other workers working in the household industry. There are 256 medium scale industries, 49084 small scale and 7500 cottage industries. There are considerable cotton and polyester hanks production in the district.

The principle agricultural crops are paddy, jowar, sugarcane, cotton, groundnut, maize, tobacco and the non-agricultural products like teak, tea, coffee, cardamom. The area is mainly irrigated by government canals and tube

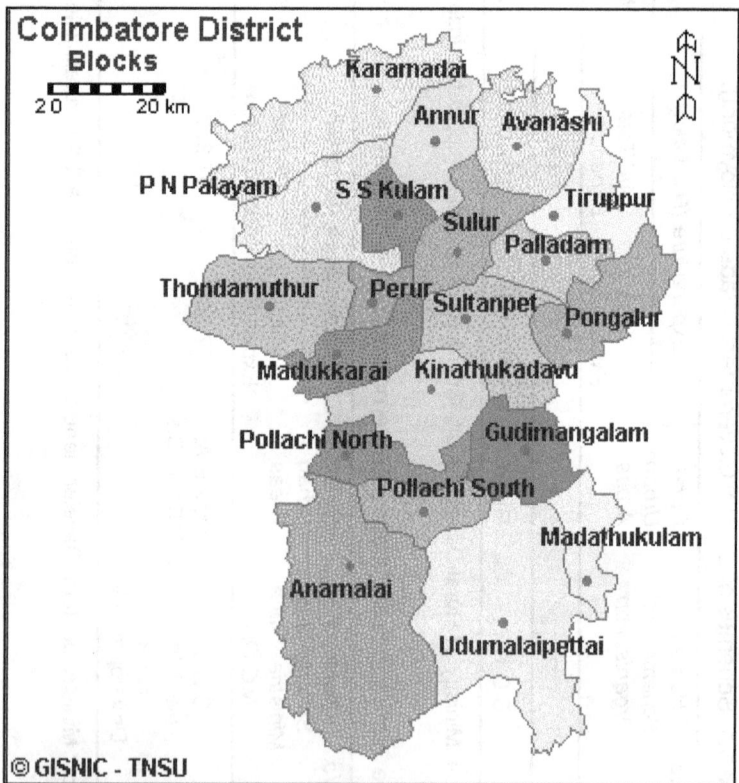

Figure 3.2: Map of Coimbatore district blocks.

wells. In some parts of the district private canals and wells are also in use. There are 15 veterinary hospitals and 47 veterinary dispensaries and one clinical centre in the district. The livestock like cattle, buffaloes and sheep are considerable in number. One dairy and 541 milk co-operative societies are there for which the SHG members sell the milk.

About 4 women welfare societies, one training centre and 8722 women development groups are there. The local bodies like corporations, municipalities, panchayat unions, town panchayats and village panchayats exist in the district. All the villages in the district are fully electrified.

Table 3.3: List of selected blocks for the study.

Tamil Nadu State					
Coimbatore District			Trichy District		
Sl.No.	Block (11)	Samples	Sl.No.	Block (19)	Samples
1.	Pollachi	3	1.	Edutiputtipudur	6
2.	Kinathukadavu	7	2.	Manikandam	5
3.	Anaimalai	2	3.	Thiruverumbur	3
4.	Karamadai	6	4.	Karumandapam	6
5.	P.N.palayam	7	5.	Andhanallur	3
6.	Udamalpet	4	6.	Lalgudi	5
7.	Gudimangalam	4	7.	Thottiyam	5
8.	Palladam	5	8.	Thurayur	4
9.	Tirupur	6	9.	Marungapuri	3
			10.	Usiri	1
			11.	Uppliyapuram	5
	Total	**44**			**46**

There are 19 blocks in the district from which 11 blocks have been selected to get the wider picture for the present study.

4

Social and Psychological Factors in Enterprise Development

4.1 Social Aspects

About 60-70 per cent of the women fall in the category of 30-50 age range who is actively involved in the entrepreneurship.

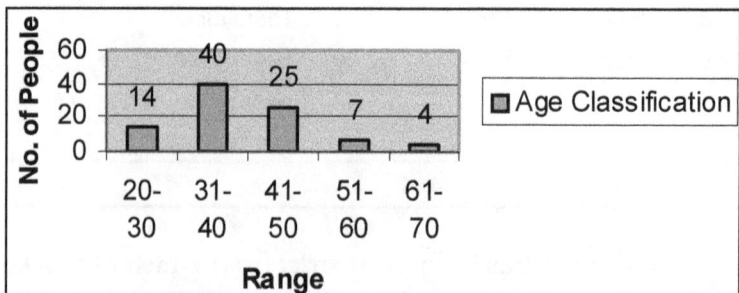

Figure 4.1: Age classification.

Married women entrepreneurs are more in number. Out of 90 women, 82 are married. The number of women in other categories is insignificant.

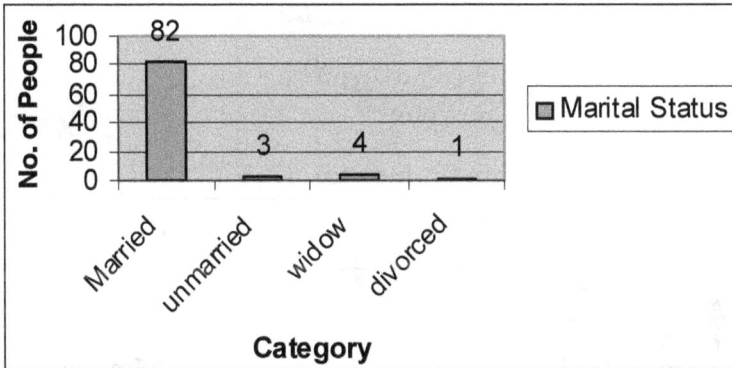

Figure 4.2: Marital status.

On an average about 50 women entrepreneurs have small family and do not have much dependents.

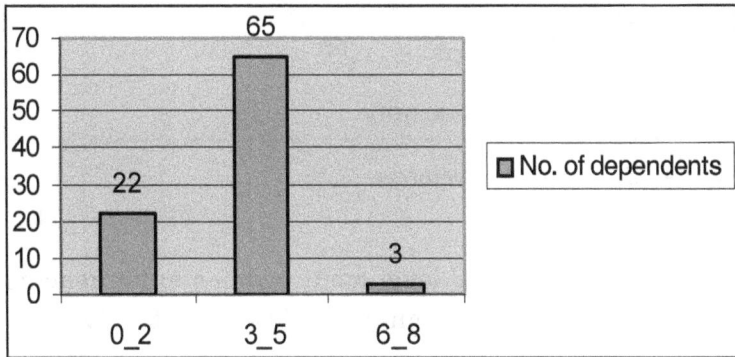

Figure 4.3: Number of dependents.

Personal income is gradually increasing. Average entrepreneurs are getting up to 10000-30000 per annum.

Bank is the major source of providing financial assistance. Around 64 entrepreneurs have received Bank assistance out of 90.

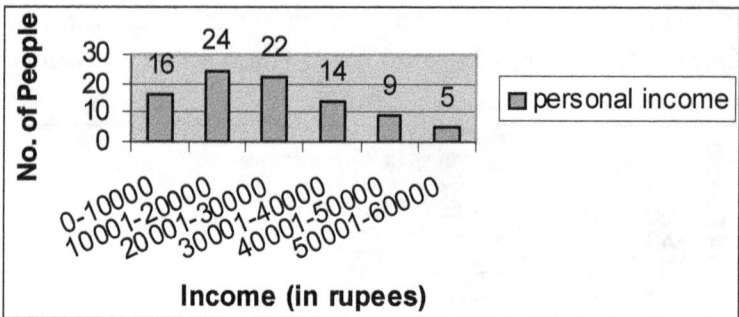

Figure 4.4: Personal income.

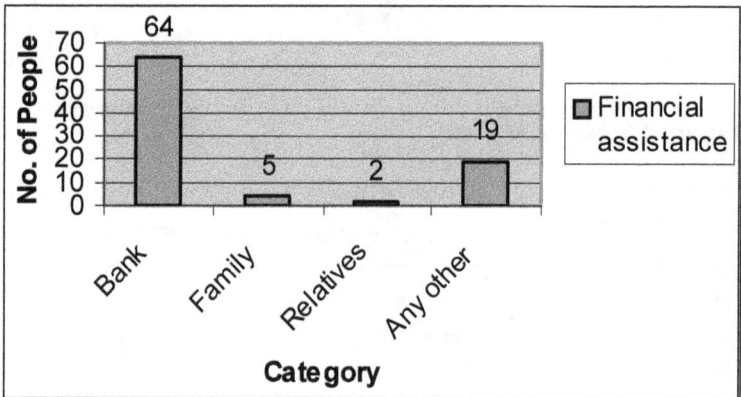

Figure 4.5: Financial assistance.

At the community level many women entrepreneurs received the highest recognition followed by family level. In other levels also a considerable number of women entrepreneurs received recognition.

Around 15 -16 women have got awards in recognition for doing good work in the area.

Growth rate is observed over a period of time based on pre and post-business growth and then gradual success over a period of time. During the work process 10 per cent growth

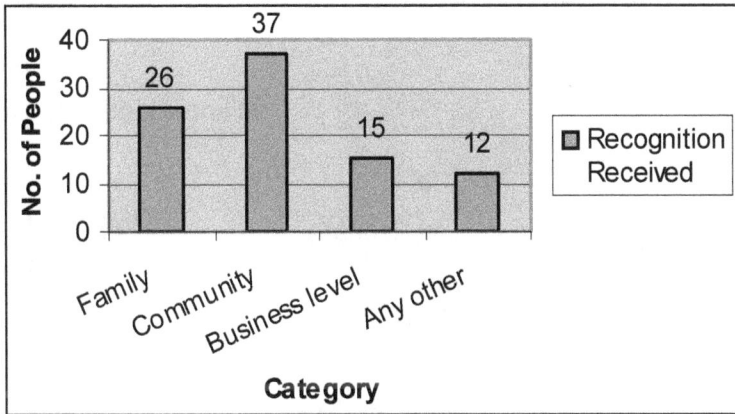

Figure 4.6: Recognition received.

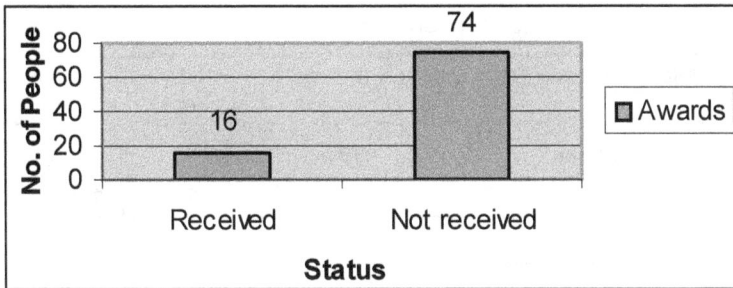

Figure 4.7: Awards received.

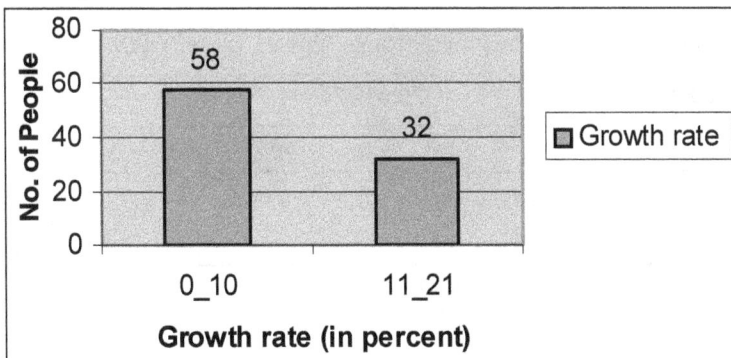

Figure 4.8: Growth rate.

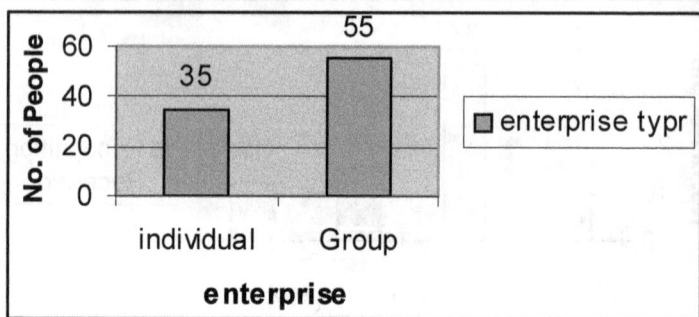

Figure 4.9: Enterprise type.

rate has been observed by around 58 rural women and the rest has observed around 20 per cent growth.

Table 4.1: Frequencies for enterprise type.

Enterprise Type		Frequency	Per cent	Valid per cent	Cumulative per cent
Valid	Individual	35	38.9	38.9	38.9
	Group	55	61.1	61.1	100.0
	Total	90	100.0	100.0	

From the Graph and the table it is clear that group members dominate the individuals with 55 out of 90.

Table 4.2: Cross tabs for enterprise type vs. growth rate.

Enterprise Type	Growth Rate						Total
	0	5	10	20	30	50	
Individual	17	6	8	4			35
Group	25	10	13	5	1	1	55
Total	42	16	21	9	1	1	90

From the cross tabulations we may conclude that there are 13 members from "group" registered 10 per cent growth and six persons with 20 per cent which are better compared to Individual enterprise. In individual enterprise 17 out of 35 registered no growth.

Table 4.3: Correlations for enterprise type and growth rate.

Correlations		Enterprise Type	Growth Rate
Enterprise type	Pearson Correlation	1	.059
	Sig. (2-tailed)	–	.582
	N	90	90
Growth rate	Pearson Correlation	0.059	1
	Sig. (2-tailed)	0.582	–
	N	90	90

Table 4.4: Frequencies for education level of entrepreneurs.

Education Level		Frequency	Per cent	Valid per cent	Cumulative per cent
Valid	illiterate	15	16.7	16.7	16.7
	Middle	42	46.7	46.7	63.3
	High School	22	24.4	24.4	87.8
	Intermediate	5	5.6	5.6	93.3
	College level	6	6.7	6.7	100.0
	Total	90	100.0	100.0	

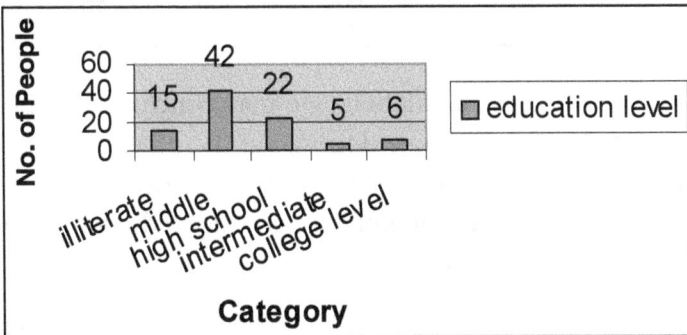

Figure 4.10: Education level.

The correlation between enterprise type and growth rate are positive with 0.059 which implies that both the variables enterprise type and the growth rate moving in the same direction. So we conclude that growth rate increases with the enterprise type and vice-versa. Group members are performing better than individuals.

Lot of entrepreneurs is from middle school background. There 42 members out of 90 are from middle school background with (46.7 per cent). After that the high school with 22 members with (24.4 per cent). The remaining is negligible. Tamil nadu being highly literacy state, rural women are educated. Only 15 per cent are illiterate and majority of them are either middle or high school pass.

Table 4.5: Cross tabs for education level vs. growth rate.

Education Level	Growth Rate						Total
	0	5	10	20	30	50	
Illiterate	9	3	3	–	–	–	15
Middle	19	7	11	4	–	1	42
High School	10	4	4	3	1	–	22
Intermediate	1	2	1	1	–	–	5
College level	3	–	2	1	–	–	6
Total	42	16	21	9	1	1	90

From the cross tabulations we may conclude that here also middle school people dominated the others, about 11 out of 42 registered (10 per cent) and 4 registered (20 per cent) growth rate. The members with high education level also doing well.

Table 4.6: Correlations for education level vs. growth rate.

Correlations		Education Level	Growth Rate
Education Level	Pearson Correlation	1	.113
	Sig. (2-tailed)	.	.290
	N	90	90
Growth rate	Pearson Correlation	.113	1
	Sig. (2-tailed)	.290	.
	N	90	90

From the table we may conclude that both the variables moves in the same direction. The correlation is positive with.113 which implies the growth rate increases with the education level and vice-versa.

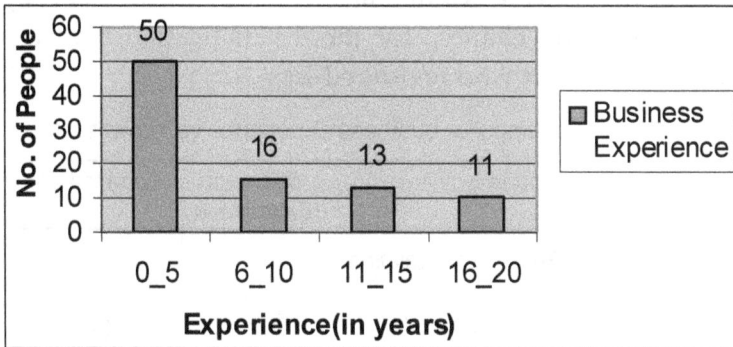

Figure 4.11: Business experience.

From the bar diagram we may conclude that there are 50 members who have the business experience between 0 to 5 years. And after that there are 11 members who have the business experience of 16 to 20 years, which is a good factor. There are about 16 people with 6 to 10 years experience.

Table 4.7: Cross tabs for business experience vs. growth rate

Business Experience	Growth Rate						
	0	5	10	20	30	50	Total
0-5	25	9	8	6	1	1	50
6-10	8	5	3	0	0	0	16
11-15	10	2	2	0	0	0	13
16-20	7	1	3	0	0	0	11
Total	50	17	16	6	1	1	90

From the table it is clearly known that the persons with 0 to 5 years experience dominate the others. There are above (50 per cent) members each from both the categories and their growth rates are comparatively good with others. They have the better chances for the diversification than the experienced ones who registered less.

Table 4.8: Correlations for business experience vs. growth rate.

Correlations		Business Experience	Growth Rate
Business Experience	Pearson Correlation	1	-.106
	Sig. (2-tailed)	.	.321
	N	90	90
Growth rate	Pearson Correlation	-.106	1
	Sig. (2-tailed)	.321	.
	N	90	90

Both the business experience and growth factor are negatively correlated with -0.106. This tells that both the variables are inversely proportional to each other and the persons with 1 to 2 years experience have better performance compared to others.

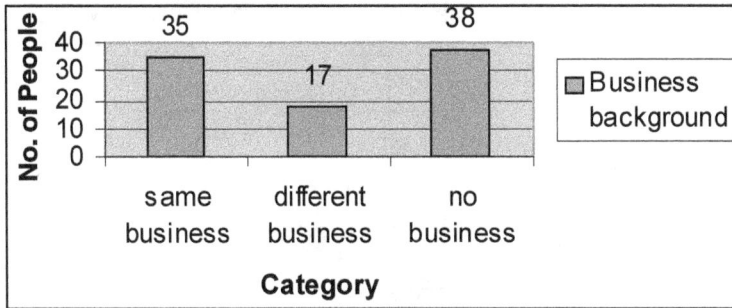

Figure 4.12: Frequencies for business background.

About 38 out of 90 members with no business background are dominates other two categories. There are 35 persons who are doing the same business and 17 doing the different business.

Table 4.9: Cross tabs for business background vs. growth rate.

Business Background	Growth Rate						Total
	0	5	10	20	30	50	
Same business	18	4	9	3	1		35
Different business	9	3	4			1	17
No business background	15	9	8	6			38
Total	42	16	21	9	1	1	90

The persons with same business background and the persons with no business background are also doing well. In "same business" background 8 registered 10 per cent and 4 registered 20 per cent. Interestingly the category "no business background" 8 registered (10 per cent) and 6 registered 20 per cent. The persons with "different business" background registered less growth. Overall the persons with same business background are doing well.

Table 4.10: Correlations for business background vs. growth rate.

Correlations		Business Background	Growth Rate
Business background	Pearson Correlation	1	.059
	Sig. (2-tailed)	.	.582
	N	90	90
Growth rate	Pearson Correlation	.059	1
	Sig. (2-tailed)	.582	.
	N	90	90

The correlation between business background and growth rate are positive. So both the variables moves in the same direction imply growth rate increases with business background and vice-versa.

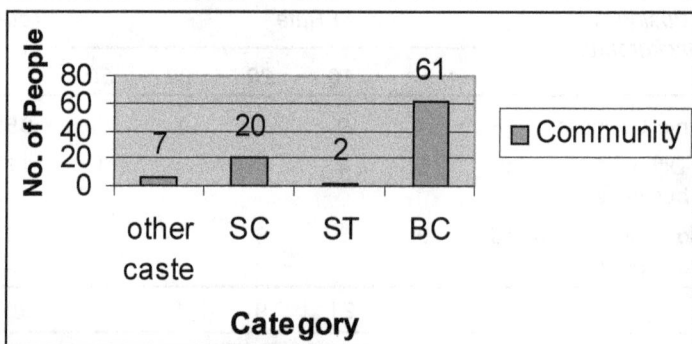

Figure 4.13: Community.

From the graph and the table it is very clear that the Backward Caste community totally dominates the other communities with 61 out of 90. After that there are 20 members from Schedule caste community (SC). Only 2 persons are there from the Schedule Tribe (ST) community

Table 4.11: Frequencies for community of entrepreneurs.

Community		Frequency	Per cent	Valid per cent	Cumulative per cent
Valid	Other caste	7	7.8	7.8	7.8
	SC	20	22.2	22.2	30.0
	ST	2	2.2	2.2	32.2
	BC	61	67.8	67.8	100.0
	Total	90	100.0	100.0	

Table 4.12: Cross tabulations for community vs. growth rate.

Community	Growth Rate						
	0	5	10	20	30	50	Total
Other caste	4	2	1				7
SC	8	5	1	5	1		20
ST	1		1				2
BC	29	9	18	4		1	61
Total	42	16	21	9	1	1	90

From the cross tabulations we understand that about (50 per cent) of the Backward community people registered some growth. There are 17 members' registered (10 per cent)

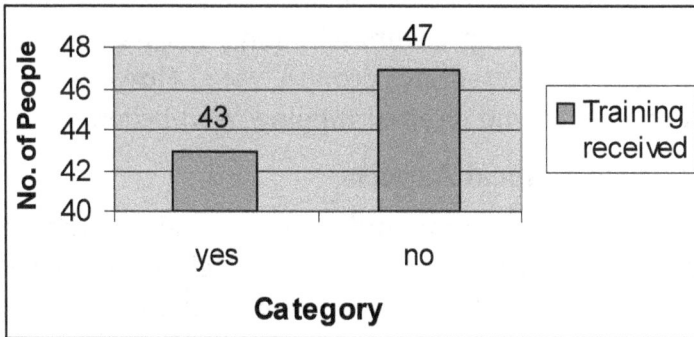

Figure 4.14: Training received.

growth and 4 registered (20 per cent). In the SC category 8 out of 20 registered (20 per cent) growth rate. Overall SC community is working well comparatively.

Table 4.13: Frequencies for training received.

		Frequency	Per cent	Valid Per cent	Cumulative Per cent
Valid	yes	43	47.8	47.8	47.8
	no	47	52.2	52.2	100.0
	Total	90	100.0	100.0	

From the table it is clear that 43 received training and 47 not received any training.

Table 4.14: Cross tabs for training received vs. growth rate.

Training received * Growth rate Cross tabulation Count							
Training Received	Growth Rate						
	0	5	10	20	30	50	Total
yes	20	6	11	4	1	1	43
no	22	10	10	5			47
Total	42	16	21	9	1	1	90

We can conclude that, from the data in the above table, 22 out of 47 entrepreneurs who have not received any training registered no growth rate. However, the entrepreneurs who received training fared better.

4.2 Psychological Aspects

The Bar diagram and the table show that 50 entrepreneurs demonstrated 'medium level' of Achievement Motivation, the highest registering 55.6 per centage. About 29 entrepreneurs demonstrated 'high-level' Achievement Motivation, while the rest 11 showed 'less level.'

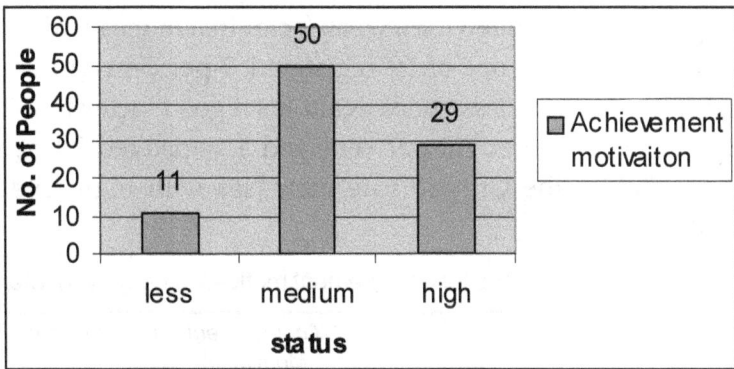

Figure 4.15: Achievement motivation

Table 4.15: Frequencies for achievement motivation.

		Frequency	Per cent	Valid Per cent	Cumulative Per cent
Valid	Less	11	12.2	12.2	12.2
	Medium	50	55.6	55.6	67.8
	High level	29	32.2	32.2	100.0
	Total	90	100.0	100.0	

Table 4.16: Cross tabs for achievement motivation vs. growth rate.

Achievement Motivation	Growth Rate						
	0	5	10	20	30	50	Total
Less	5	3	2	1			11
Medium	26	8	13	2		1	50
High level	11	5	6	6	1		29
Total	42	16	21	9	1	1	90

From the table we conclude that the achievement motivation at high level is registered the good figures with 6 out of 29 registered 10 per cent growth and 7 registered

20 per cent. It is somewhat a weak factor that in the medium level category 26 out of 50 registered 0 per cent growth rates. Coming to "less" once again it's a good factor that 2 out of 11 registered 10 per cent and 1 registered 20 per cent. Finally the Growth rate increases with high level motivation.

Table 4.17: Correlations for achievement motivation vs. growth rate.

		Achievement Motivation	Growth Rate
Achievement motivation	Pearson Correlation	1	.136
	Sig. (2-tailed)	.	.201
	N	90	90
Growth rate	Pearson Correlation	.136	1
	Sig. (2-tailed)	.201	.
	N	90	90

The correlation between achievement motivation and the growth rate is positive with 0.136. So we may conclude that both the variables move in the same direction. If the

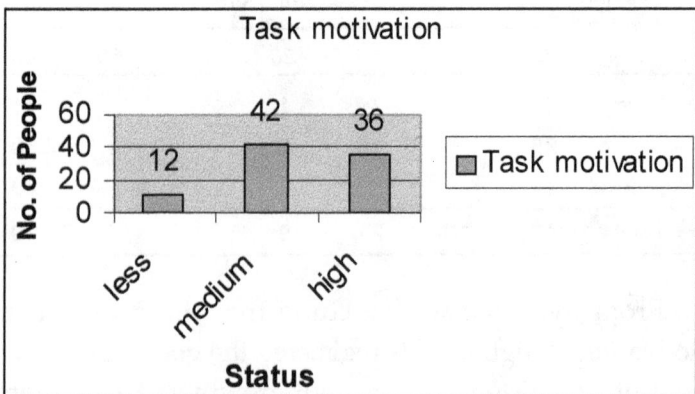

Figure 4.16: Task motivation.

achievement motivation is high then growth rate is high and vice-versa.

Table 4.18: Frequencies for task motivation.

Task Motivation		Frequency	Per cent	Valid Per cent	Cumulative Per cent
Valid	Less	12	13.3	13.3	13.3
	Medium	42	46.7	46.7	60.0
	High level	36	40.0	40.0	100.0
	Total	90	100.0	100.0	

From the table we may conclude that 42 members has medium level of task motivation and 36 members have the high level of task motivation and only 12 members have less level of task motivation. The frequency is high in Medium and high level task motivation.

Table 4.19: Cross tabs for task motivation vs. growth rate.

Task Motivation	Growth Rate						
	0	5	10	20	30	50	Total
Less	6	2	2	2			12
Medium	21	8	10	2		1	42
High level	15	6	9	5	1		36
Total	42	16	21	9	1	1	90

From the cross tabulations we may know that, 9 out of 42 from medium level task motivation registered 10 per cent growth rate and 2 registered 20 per cent. It is interesting that about 6 members out of 36 registered 20 per cent growth rate from high level task motivation. The frequency is more in medium and high level task motivation.

Table 4.20: Correlations for task motivation vs. growth rate.

Correlations		Task Motivation	Growth Rate
Task motivation	Pearson Correlation	1	.054
	Sig. (2-tailed)	.	.615
	N	90	90
Growth rate	Pearson Correlation	.054	1
	Sig. (2-tailed)	.615	.
	N	90	90

From the table we may conclude that the relation between task motivation and the growth rate is positive. It implies that both the variables move in the same direction and growth rate increases with the task motivation and vice versa. The growth rate increases with medium and high level task motivation.

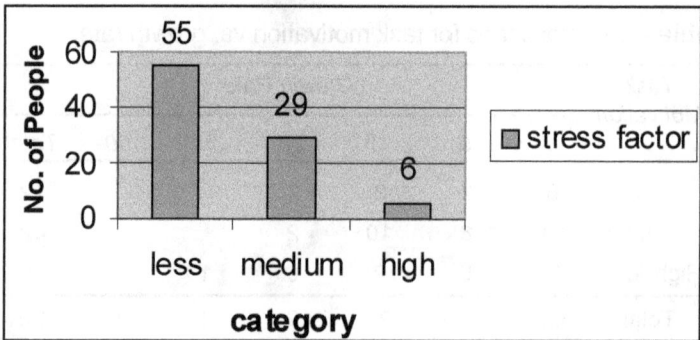

Figure 4.17: Stress factor.

It is a good factor that lot of members feels "less stress "with multiple roles. There are 55 members in less stress category out of 90. After that 29 members fall in medium level stress and only 6 members have "high level" stress. As per the stress findings moderate stress shows higher productivity.

Table 4.21: Frequencies for stress factor.

Stress Factor		Frequency	Per cent	Valid Per cent	Cumulative Per cent
Valid	Less	55	61.1	61.1	61.1
	Medium	29	32.2	32.2	93.3
	High level	6	6.7	6.7	100.0
	Total	90	100.0	100.0	

Table 4.22: Cross tabs stress factor vs. growth rate.

Stress Factor	Growth Rate						
	0	5	10	20	30	50	Total
Less	25	8	14	6	1	1	55
Medium	13	8	6	2			29
High level	4		1	1			6
Total	42	16	21	9	1	1	90

It is strange that from less stress category 25 out of 55 registered 0 per cent growth rate and 13 out of 29 from medium level category also registered 0 per cent growth rate. The persons with high level and medium level stress are doing comparatively well.

Table 4.23: Correlations for stress factor vs. growth rate

Correlations		Stress Factor	Growth Rate
Stress Factor	Pearson Correlation	1	-.102
	Sig. (2-tailed)	.	.336
	N	90	90
Growth Rate	Pearson Correlation	-.102	1
	Sig. (2-tailed)	.336	.
	N	90	90

From the table we may conclude that the correlation between Stress factor and growth rate are negative it implies that stress factor and the growth factor are inversely proportional to each other. There is no dependency between stress factor and growth rate.

Table 4.24: Correlations for training received vs. stress factor

Correlations		Training Received	Stress Factor
Training received	Pearson Correlation	1.000	-.195
	Sig. (2-tailed)	.	.066
	N	90	90
Stress factor	Pearson Correlation	-.195	1.000
	Sig. (2-tailed)	.066	.
	N	90	90

From the table, we may conclude that the correlation between Training Received and locus of control is negative. And we also conclude that both the variables are moving in the opposite direction *i.e.* the more training received in

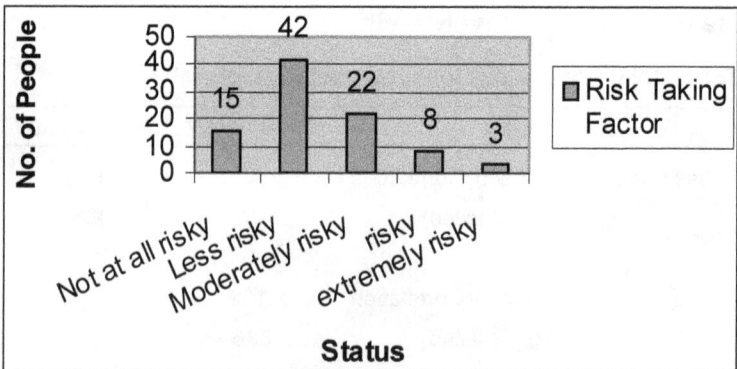

Figure 4.18: Risk taking behaviour.

the area of their business the less stress is perceived. Training was given to enhance the knowledge, skill and attitude.

Table 4.25: Frequencies for risk taking behaviour.

		Frequency	Per cent	Valid Per cent	Cumulative Per cent
Valid	Not at all risky	15	16.7	16.7	16.7
	Less risky	42	46.7	46.7	63.3
	Moderately risky	22	24.4	24.4	87.8
	Risky	8	8.9	8.9	96.7
	Extremely risky	3	3.3	3.3	100.0
	Total	90	100.0	100.0	

The table shows that the 'risk taking behaviour' of the entrepreneurs. About 42 out of 90 fall in "less risky" category and the per centage works out to be 46.7. the next highest category is "moderately risky" with 22 entrepreneurs with 24.4 percentage. About 15 entrepreneurs with 16.77 percentage fall in 'not at all risky' category, very few fall under 'risky (8) and 'extremely risky' (3) categories.

Table 4.26: Cross tabs for risk taking behaviour vs. growth rate.

Risk Taking Behaviour	Growth Rate						
	0	5	10	20	30	50	Total
Not at all risky	10	3	1	1			15
Less risky	15	6	13	7		1	42
Moderately risky	8	7	6	1			22
Risky	8						8
Extremely risky	1		1		1		3
Total	42	16	21	9	1	1	90

From the cross tabulations it is understood that the persons with "less risky" behaviour registered good growth rates with 12 members registering 10 per cent and 7 registering 20 per cent and the persons with "moderately risky" also registered good growth rate.

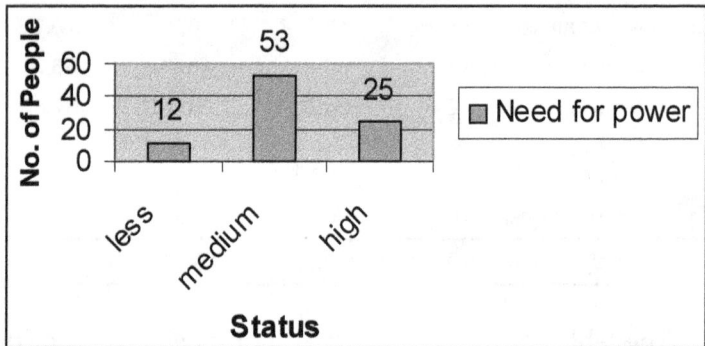

Figure 4.19: Need for power.

About 50 per cent people expect power in their business circle, as it reflects responsibility and attitude to do more in comparison to other entrepreneurs.

Table 4.27: Frequencies for need for power.

		Frequency	Per cent	Valid Per cent	Cumulative Per cent
Valid	Less	12	13.3	13.3	13.3
	Medium	53	58.9	58.9	72.2
	High level	25	27.8	27.8	100.0
	Total	90	100.0	100.0	

The table shows that the need for power. The highest number (53) falls under 'medium level' with 58.9 percentages followed by 'high level' and 'less level' (12).

Table 4.28: Cross tabs for need for power vs. growth rate.

Need for Power	Growth Rate						
	0	5	10	20	30	50	Total
Less	6	3	1	2			12
Medium	23	9	16	3	1	1	53
High level	13	4	4	4			25
Total	42	16	21	9	1	1	90

From the cross tabulations we may conclude that there are 15 members who registered 10 per cent growth rate and 4 registered 20 per cent who needs medium level power. The members with high level need for power also registered good growth rate.

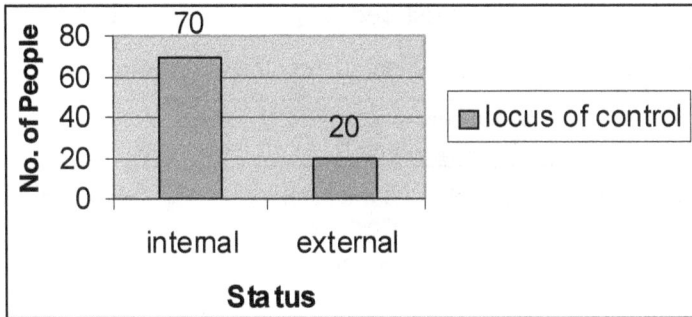

Figure 4.20: Locus of control.

Table 4.29: Frequencies for locus of control.

		Frequency	Per cent	Valid Per cent	Cumulative Per cent
Valid	Internal	70	77.8	77.8	77.8
	External	20	22.2	22.2	100.0
	Total	90	100.0	100.0	

From the graph and the table we may conclude that the about 80 per cent of the members have locus of control as internal and only 20 from external factor.

Table 4.30: Cross tabs for locus of control vs. growth rate.

Locus of Control	Growth Rate						
	0	5	10	20	30	50	Total
Internal	34	12	14	8	1	1	70
External	8	4	7	1			20
Total	42	16	21	9	1	1	90

From the cross tabulations it is clear that 34 out of 70 from internal category registered 0 per cent growth rate, that is about 50 per cent which is very poor. And only 9 members out of 70 registered 20 per cent growth rate.

Table 4.31: Correlations for locus of control vs. growth rate.

Correlations			Locus of Control	Growth Rate
Kendall's tau_b	Locus of Control	Correlation Coefficient	1.000	.031
		Sig. (2-tailed)	.	.748
		N	90	90
	Growth Rate	Correlation Coefficient	.031	1.000
		Sig. (2-tailed)	.748	.
		N	90	90

The correlation between locus of control and the growth rate is positive, which implies that both the variables moving in the same direction. So the inference is if locus of control is internal the growth rate increases and vice-versa. In this

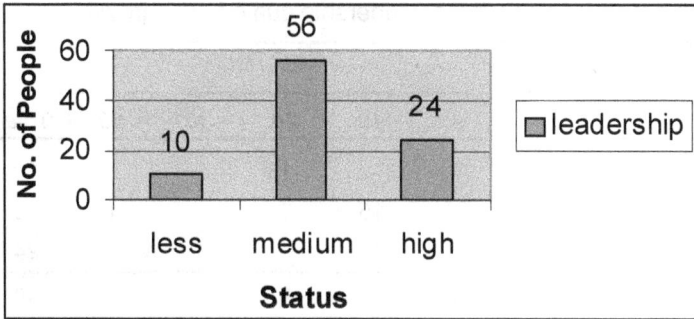

Figure 4.21: Leadership qualities.

particular case we applied Kendal's tau_b test instead of Pearson's test.

Table 4.32: Frequencies for leadership qualities.

		Frequency	Per cent	Valid Per cent	Cumulative Per cent
Valid	Less	10	11.1	11.1	11.1
	Medium	56	62.2	62.2	73.3
	High level	24	26.7	26.7	100.0
	Total	90	100.0	100.0	

From the table it is understood that maximum number of the entrepreneurs has the medium level of leadership qualities with 56 out of 90 falls in this category. About 24 have high level leadership qualities and only 10 have less leadership qualities.

From the cross tables we may conclude that from medium level category 14 registered 10 per cent and 3 registered 20 per cent growth rate. From high level also 6 each registered 10 per cent and 20 per cent respectively.

Table 4.33: Cross tabs for leadership qualities vs. growth rate.

Leadership Qualities	Growth Rate						
	0	5	10	20	30	50	Total
Less	6	3		1			10
Medium	25	12	15	3		1	56
High level	11	1	6	5	1		24
Total	42	16	21	9	1	1	90

Table 4.34: Correlations for leadership qualities vs. growth rate

Correlations		Leadership Qualities	Growth Rate
Leadership Qualities	Pearson Correlation	1	.155
	Sig. (2-tailed)	.	.146
	N	90	90
Growth Rate	Pearson Correlation	.155	1
	Sig. (2-tailed)	.146	.
	N	90	90

From the above table we conclude that there is a positive correlation between leadership qualities and the growth rate. Both the variables move in the same direction with correlation.155. So the inference is growth rate increases when leadership qualities increase and vice-versa.

The correlation between training received and leadership qualities are positive and it shows that those who take training have more leadership qualities than who do not take any training.

Table 4.35: Correlations for training received and leadership qualities.

Correlations		Training Received	Leadership Qualities
Training Received	Pearson Correlation	1.000	.236
	Sig. (2-tailed)	.	.025
	N	90	90
Leadership Qualities	Pearson Correlation	.236	1.000
	Sig. (2-tailed)	.025	.
	N	90	90

* Correlation is significant at the 0.05 level (2-tailed).

5

Case Studies

In depth case studies of 90 successful rural women entrepreneurs were undertaken. A detailed understanding of the close environment of these rural women; and their connectivity and relationship with their families and community; and their enterprises, hindrances and facilitating factors were collected through these case studies. The systematic pattern comes out clearly the differences of approach, attitudes, aspirations, motivation and support systems of these entrepreneurs. During the case studies entrepreneurs' development and growth were recorded systematically. Complete profiles of about 41 entrepreneurs have been included hereunder as sample.

5.1 Case Study 1

Kalavathi, a 45 year old woman entrepreneur, hails from Vakkampalayam Panchayat, Pollachi Taluka in

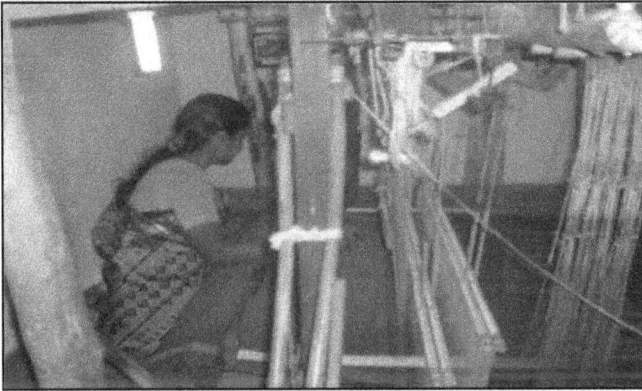

Figure 5.1: A view of handloom weaving.

Coimbatore district. She is into handloom weaving. She is from a backward class community and a middle school educated.

She joined "Sri Roja Self Help Group", which helped her getting financial help to grow her business. Hers is a family of three. Her husband helps her in her business. He helps her in purchasing raw material and in selling her products, sarees.

She was born into a poor family and her parents were illiterate. They are daily wage labourers, who live in a joint family.

Having no business background, Kalavathi ventured into business with Rs.15,000/-, as capital investment. She has been trained in embroidery and doll-making. She is also being given some computer training organised by her Self Help Group with the help of the Panchayat.

She is playing the roles of a mother and a business woman. She has a medium level achievement motivation and her locus of control is internal. Her leadership qualities and task motivation, the both, are of middle level. She exhibited a moderate risk taking behaviour.

5.2 Case Study 2

Subbammal (56) hails from Arasampalayam, Pollachi block of Coimbatore district. Being an illiterate backward class woman, she started her career as a daily wage labourer.

After joining a Self Help Group, she started her own handloom weaving business with a capital investment of Rs.25,000/- obtained from Bank of Baroda as a loan.

Her illiterate parents are also into handloom weaving. Their annual income is about Rs.18,000/-. Because of high esteem for women in their family, women are very active in decision-making in the family.

Subbammal started her business after obtaining a formal training of 15 days from the Self Help Group. She employed about 12 workers and her business assets worth about Rs.10,000/-. Since her husband extends all financial and marketing support to her, she looks after the production. She has good leadership skills and feels her business less risky.

Despite little discouragement from her family in the initial stages, she succeeded in becoming a proven entrepreneur. She attracted her business by allowing maximum discounts on selling, especially school uniforms. She does a brisk business during festival seasons, Diwali and Pongal.

The Self Help Group members and her husband as well extend every kind of support in case of any problem.

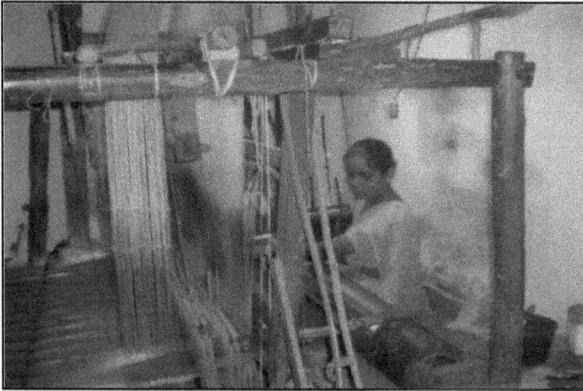

Figure 5.2: A view of handloom weaving.

5.3 Case Study 3

Jyothimani, a 31 year old backward class high-school educated woman entrepreneur is from Vakkapalayam in Pollachi block of Coimbatore district. She ventured into her own handloom weaving production after having undertaken job works earlier. Her illiterate parents are also into the same profession.

She started her business with Rs.25,000/-, financial assistance extended through State Bank of India, Pollachi Branch. Her annual income is about Rs.24,000/-, which she gets from producing and selling sarees.

She registered herself as a member of an Self Help Group on 31-12-2004. The Panchayat president is one among these SHG members. About 70 per cent of her villagers are educated. She underwent formal education (literacy) training through "Valar Kalvi" scheme promoted by the Government for senior citizens and school dropout children.

5.4 Case Study 4

S. Jayalakshmi, a college dropout backward class

Figure 5.3: A view of muruku and file tags production.

woman stands as a bold statement of entrepreneurship. This 48 year old entrepreneur is from Edamalapatty Pudur in Tiruchirapalli. She has two businesses – one 'muruku' production, the other 'file tags' production. Her annual income is about Rs.24,000/-.

She is from a joint family in which women's position is active and they take all decisions in the family. Her father is a high school dropout. He works in a canteen. Their annual income is about Rs.60,000/-.

Jayalakshmi has set up a small 'file tag' making unit nearby her house. In fact, she learnt file tag making to help her son who runs a unit. Later on, she established her unit. She recruited about 40 people to work in her unit.

A small hand machine that costs about Rs.350/-, is used in making these tags. The machine is used to cut small pieces of iron sheet and fix them to the both ends of a tag. She makes Rs.1 profit out of 100 tags.

Jayalakshmi was provided Rs.7, 500/- financial assistance by Five Star SHG in which she is a member. She has no knowledge about the end users of the file tags. She merely supplies them to the dealers. She is decision maker in the family. She opines that economic independence makes women courageous and world-wise, but laments that there is less employment potentiality for uneducated rural women.

5. 5 Case Study 5

Maheshwari, a 35 year old entrepreneur lives with her potter husband Arasampalayam, Kimathukadavu block in Coimbatore district. She is also from a family of potters.

She was instrumental in the formation of a Self Help Group named as 'Seventhi Self Help Group' with the assistance of Coimbatore Multiple Service (CMS). The government trained the members of the group. The CMS extends Rs.10,000/- financial assistance to Maheshwari as a capital investment to start a pottery.

Figure 5.4: A view of making of pots.

Now, her husband makes the pots and she sells them in the market. Their income is about Rs.30,000/- per annum. Her business is usually unstable due to seasonal fluctuations. During rainy season, it is not possible for potters to make pots. In such seasonal gaps, her family does not get any income but the Seventhi SHG comes to her rescue and provides financial assistance in case of dire need.

She commands good respect and support in her family circle. Her social standing as a woman entrepreneur is positive. Her achievement motivation is medium and the locus of control is internal. In spite of being a mother and an entrepreneur, her stress levels are low. Her leadership qualities and the need for power are medium. Her task motivation is less and she has a risky behaviour.

5.6 Case Study 6

Saraswathi, a 35 year old entrepreneur is from Nallarajan Colony, Edamalapattipudur in Tiruchirapalli district. She lives with her husband and two sons. This backward class middle school dropout is into *'papad'* making. Her annual income is about Rs.3600/-.

She is from a nuclear family and her parents also had middle school level education. Women's position is secondary in their family. Their annual income is about Rs.36,000/-.

Because of their abject poverty, Saraswathi's elder son works as a mechanic. She started *'papad'* making to supplement family income. Her husband helps her in marketing the product. They have to depend on small grocery shops for selling the product. Supermarket managements do not buy such product because it does not have a brand name. So, Saraswathi is very keen on having her product registered.

Saraswathi started her business with a capital investment of Rs.20,000/- borrowed from her parents. She has not registered her unit so far, despite being in the business for about 10 years.

5.7 Case Study 7

Lakshmi, a 36 year old entrepreneur hails from Kovilpalayam Panchayat of Kinathakadavu blocks in Coimbatore district. She is from middle class and educated up to middle school level.

She is the founder member of Kalpana Chawla Self Help Group. She started handloom business with a capital investment of Rs.60,000/- which she sought from Bank of Baroda as a loan. She was trained to run the business by a handloom society for a year. Net assets of her business now worth about Rs.80,000/-. Her husband helps her in managing the enterprise. Altogether about 12 members of her family work for her.

Being a successful entrepreneur, she is hailed as a 'role model'. She helps other members of the group in earning more while she herself earns about Rs.6,000/- per annum. All her family supports her in all aspects and she is happy with her business.

She has high achievement motivation and external locus of the control. She demonstrates medium levels of stress despite playing multiple roles. She has medium levels of leadership qualities as well as the need for power. She has less risk taking behaviour and medium level task motivation.

5.8 Case Study 8

S. Rajaveni (43) a most backward caste woman, is from Arasampalayam in Kinthukadavu block of Coimbatore district. She dropped out of her school after 6th standard. Only four among 12 members of her family are dependents.

She has two sons and daughters. Sanginthala, eldest of the daughters, is in M.Sc., and the other one Anandhi is

Figure 5.5: A view of tea and coffee stall.

in M.A. The eldest of the sons, Selvaraj is an arts graduate while the other one, Kangathavan is in B.Sc., Women's position in the family is secondary. Members of the family take all important decisions jointly.

Rajaveni started her career as a daily wage worker before setting up a tea and coffee stall in her house. She also used to supply public address systems on hire. Gradually her tea stall business grew into a motel (mess). Her capital investment accounted for Rs.25,000/-.

Her husband and other members of her family helped her run the business successfully. Being a member of an SHG, she participated in many social programmes such as 'Awareness on Education and Child Labour, Women Development and the programme, Manitha Chayili (a programme meant for educating women to work and to become entrepreneurs).

While her achievement motivation is medium, her locus of control is medium. She has less stress levels and power. Her task motivation is less.

5.9 Case Study 9

Thirumal is a 45 year old scheduled caste woman entrepreneur from Somayanpalayam in Kinathukadavu block in Coimbatore district. She studied upto middle school level.

She is from a joint family and her parents are also educated upto middle school level. She worked as an agricultural labourer on daily wages, before joining an SHG initiated by an NGO, Coimbatore Multiple Social Service Society (CMSS). Her group is into producing dairy products. Each member of her group contributed Rs.3,000/- towards initial investment and CMSS provided Rs.2000/- each additionally. In spite of being started 20 years ago, the SHG has been producing into full capacity only for about 6 years.

Thirumal's husband is a daily wage worker. Being a woman and being a scheduled caste woman especially, she had to face many problems initially. Her own family, mainly the male members, scorned at her but after her success, they started extending their own support. Her group, own, stands a 'role model' among other groups of neighbouring villages.

Thirumal has medium level of leadership and she feels she manages this business with all the members' support. Though she plays multiple roles in her family and in group she doesn't feel any stress. The achievement motivation is medium and the locus of the control is internal. The stress associated with the multiple roles is less. The leadership qualities are medium and the need for power is medium. The task motivation is less. Risk taking behaviour is risky.

Figure 5.6: A view of petty shop.

5.10 Case Study 10

A, Maliga, a scheduled caste woman entrepreneur hails from Peria Milaguparrai Panchayat, Vaduvar Street in Tiruchirapalli district. This 43 year woman studied up to middle school level. She is a member of Malligai Self Help Group. She has to daughters and a son. Her husband works as a labourer in a Government Office.

Maliga started a petty shop in her own house with an investment of Rs.10,000/- given by her husband, inorder to support her debt ridden family.

She is from a joint family. Her parents are educated and their annual income is about Rs.72,000/-. All the important decisions are taken by men in the family. The women's position in the family is secondary.

5.11 Case Study 11

T. Rathna, a scheduled caste entrepreneur turned daily wage earner is from Thamarikulam in Kinthukadavu block of Coimbatore district. This 5th standard educated woman has been into coir-rope manufacturing business since 1999.

Figure 5.7: A view of coir rope making.

Rathna's parents are also middle-school educated and are daily wage earners. Women are highly esteemed in their family and decisions in the family are left to them. Their annual income is about Rs.1, 10,000/-.

She is a member of Indira Gandhi Self Help Group. The group started the coir-role business with a capital investment of Rs.25,000 and Bank of Baroda provided Rs.60,000/- loan. Ratna manages 15 members of the group. Tamil Nadu government donated mobile phones to each member of the group.

The group markets part of their produce locally but the major part is sold in Coimbatore. They make 20 paisa profit on each role. The group is now looking for help from the government to procure a large premises as well as financial support for buying machinery.

Ratna has good knowledge of rope-making business and trains others too. She has medium level of leadership qualities and never feels stress in her work. She is ready to play multiple roles and loves to take less risk so that it is manageable for her and her group.

5.12 Case Study 12

V. Kittal, an illiterate scheduled caste woman is an entrepreneur in Solavampalayam village in Kinathukadavu taluk of Coimbatore district. They are a family of 2 and only one is dependent.

She is a member of Thirumal Self Help Group. She is into building works. Her personal income is Rs.18,000/- per annum. She was born into a nuclear family. Her illiterate parents earn their bread working as daily wage labourers. Their annual family income accounts to about Rs.15,000/-. Women's position in the family is active and they are the decision makers in the family. They are into handloom business.

Kittal registered her proprietary enterprise on 16th February 2007 with a capital investment of Rs.5,00,000/- borrowed from Bank of Baroda. She employed 12 people and the net worth of her enterprise assets is Rs.1, 50,000/-. She works from 9:30 in the morning till 6 O' clock in the evening. She looks after every aspect of her business by herself.

Her achievement motivation is medium and the locus of control is internal. She undergoes medium levels of stress. Her need for power and task motivation is medium while risk taking behaviour is moderate.

5.13 Case Study 13

Jaibureesha is an enterprising woman. This 34 year old high school dropout backward class entrepreneur hails from Voda Chittore village in Kinathukadavu block of Coimbatore district. Being a hard and skilled person, she appears to have born instincts to help others and succeed

Figure 5.8: A view of art of tailoring.

in everything she does. They are a family of four with two dependents.

Jaibureesha's husband encouraged her initially buying a sewing machine for her. She learnt the art of tailoring and soon started teaching the skills to her neighbours. She wanted her neighbours to have a better livelihood. Having such background, she established a garment unit, Raja Kootham Mahalir Self Help Group with the help of an NGO 10 years ago. She trained every member of her group. All of them are skilled and work hard. There are 3 other workers who are not part of the group. The group initially started their enterprise with Rs.25,000/- as Revolving Fund. Upon repayment of the fund, the group was sanctioned Rs.5 lakh by DRDA, Coimbatore under SGSY scheme.

The group has won several accolades and awards including 'Bharat Nirman Public Awareness Campaign' from MOI and BGOI. All business decisions are taken by the group collectively. Individually, the group members play multiple roles of being mothers, workers and care takers.

5.14 Case Study 14

K.C. Mani Mekalai is a 48 year old entrepreneur from Ezhil Nagar in Thirunumbur block in Tiruchirapalli district. She used to be a tailor, stitching clothes for women for small amounts of money. She enrolled as a member in a Self Help Group. The Self Help Group was sanctioned Rs.2, 50,000/ - to start garment business. It was then she was nominated as a manager to look after the business.

She is from a nuclear family. Her father is an employee in BHEL and his annual income is about Rs.1, 20,000/-. Women command respect in the family and the decisions in the family are taken jointly.

There are three women employees in the garment unit managed by Mani Mekalai. These employees were trained by the Government of Tamil Nadu. Mani Mekalai is trying to repay the loan as early as possible so that she can obtain more loan to expand the business. She dreams of becoming an owner of a garment industry by herself. The value of assets of her industry amounts to about Rs.2,80,000/-.

Figure 5.9: A view of stitching clothes for women.

Figure 5.10: A view of manufacturing of plastic bottles.

5.15 Case Study 15

Lakshmi, a 30 year old backward class entrepreneur hails from Subbegounderpudur village in Anaimalai block of Coimbatore district. She is a dropout from middle school. They are a family of four.

Lakshmi is a member and co-ordinator of Rajiv Gandhi Self Help Group. The Self Help Group was formed on 2nd November, 2000 with 12 members. Before joining the SHG, Lakshmi was into coir mat making. Lakshmi is from a nuclear family. Her parents are also educated upto middle school level. They are into plastic molding. Women have high esteem in their family and the decisions in the family are left to them.

Rajiv Gandhi Self Help Group manufactures plastic bottles. Their capital investment is about Rs.2, 50,000. Financial assistance was extended by Canara Bank. They have excellent track record and enormous growth. They are now planning to produce plastic bottles and seal cut products.

The SHG sells their products in 29 villages around them. The group participated in many exhibitions including the ones help at Mahatma Gandhi Stadium and Agricultural University. Few NGOs support the SHG in all aspects. The SHG members train likely entrepreneurs of other districts of Tamil Nadu and of other states such as Kerala, Karnataka as well.

The Self Help Group was a recipient of many awards, such as the Best Self Help Group Award (2005), Manimeghalai Award twice and an Rs.10,000/- cash award.

5. 16 Case Study 16

Sushila (45) is a backward class woman from Odanthurai village of Karamadai block in Coimbatore district. They are a family of eight.

The entrepreneur turned a daily wage agricultural worker ventured into packaged drinking water industry. She earns Rs. 1, 40,000/- annually. She was born into a joint family. Her illiterate parents are daily wage workers. The women's position in their family is secondary but the family decisions are taken jointly. Her parents annual income is about Rs.36,000/-.

Sushila is a member of a co-operative unit (federation). Their federation was initiated on 26th November, 2006, with a capital investment of Rs.2,00,000/-. The financial assistance was provided by Central Bank of India. The necessary training was provided to all 23 members of the federation by the government. The World Bank team visited the federation and gave suggestions improving the business.

Sushila's personal income amounts to Rs.300 in a week. She has medium level achievement motivation and external

Figure 5.11: A view of packaged drinking water.

locus of control. Being helped by her husband, she feels less stress. Her leadership qualities are medium. Her need for power is less but task motivation is high. The level of her risk taking behaviour is very less.

5.17 Case Study 17

Krishnaveni, a 35 year old scheduled caste woman is from Samanthuvapuram village in Karumandapam block in Tiruchirapalli district. She studied Intermediate. She has set up a small vegetable shop. Her annual income is about Rs.58,000/-. They are a family of four.

She is from a nuclear family. Her father is a graduate. His income is about Rs.60,000/- per annum. Decisions are

Figure 5.12: A view of small vegetable shop.

taken jointly and the women's position in the family is secondary.

Krishnaveni is a member of Durgadevi Self Help Group. She was provided loan from Indian Bank. She was also trained in tailoring at NIFT through a government sponsored programme. She also does tailoring in addition to selling vegetables. There has been a steady 25 per cent growth in her business. Her business assets are worth about Rs.13,000/-.

5.18 Case Study 18

Parvathi is a 50 years old illiterate backward class entrepreneur living in Odamthurai village near Methupalayam in Karamadai taluk of Coimbatore district. She has a family of six.

Parvathi was born to illiterate parents in a joint family. Her parents are daily wage agricultural workers. Women have secondary status in their family but decisions are taken jointly. Their annual family income is about Rs.36,000/-.

Parvathi and her husband were working as daily wage workers in a tea plant. She joined a Self Help Group about

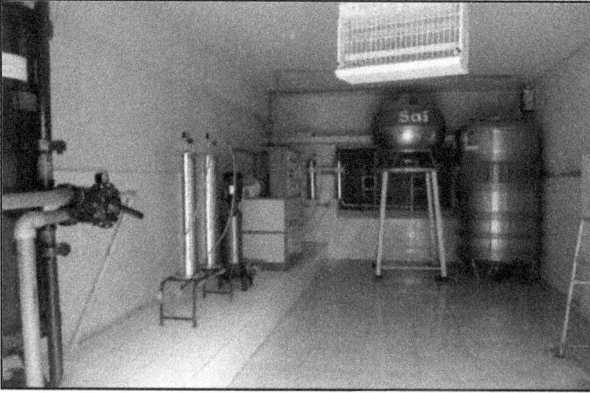

Figure 5.13: A view of packaged drinking water plant.

one and a half year ago. Her annual personal income amounts to Rs.9, 360/-.

The SHG, in which she is a member, is into packaged drinking water business. The group was registered on 26th November 2006. Their capital investment is about Rs.20 lakhs, obtained from Central Bank of India as a loan. The World Bank team visited the unit and suggested improvements.

5.19 Case Study 19

V. Shantamani hails from Knaramkundru village of Karamadai taluk in Coimbatore district. This 33 year old backward class entrepreneur is high school drop out. Theirs is a family of six.

Shantamani used to earn a living through animal husbandry. Her high school educated parents work as daily wage earners in agriculture sector. Their income is about Rs.1 lakh per annum. They were into poultry farming in the past. Theirs is a joint family in which women's position is neutral but the family decisions are taken jointly.

Shantamani joined as a member in a Self Help Group, the group is into packaged drinking water business. The group started the business about one and a half month ago with a capital investment of Rs. 5 lakh out of which Rs.4.25 lakh were obtained from a bank as a loan and Rs.25,000/- from DRDA. All the 12 members of the group have been provided training. The asset value of the unit is about Rs.3.5 lakh. The group takes care of entire management.

5.20 Case Study 20

K. Jayalakshmi, a 35 year old entrepreneur hails from Annalai village of Andhanallur block in Tiruchirapalli district. She belongs to backward class community. She is a middle school dropout. Her husband passed away leaving her and two sons. She used to be a daily wage earner before entering into an agricultural allied enterprise.

She is from a nuclear family. Her parents are educated and they are daily wage earners. Women's position in their family is active and they take all important decisions in the family.

Her husband's death stood as an inhibiting factor in her life. Society gave her considerable support. She

Figure 5.14: A view of agricultural allied enterprise.

welcomes new entrepreneurs and encourages them to start new lives. She feels that mutual understanding among group members as well as their hard work brings success in any business.

Her achievement motivation is medium and the locus of control is internal. The stress associated with the multiple roles is medium. Her leadership qualities and the need for power are of high levels. She has high level of task motivation. Her risk taking behaviour is extremely risky.

5.21 Case Study 21

Rukmini a backward class illiterate entrepreneur is from Thenpenmudi village in Karamadail taluk of Coimbatore district. The 65 year old woman has 40 years of experience in handloom weaving. There are 9 members in her family.

She is from a nuclear family. Her high school educated parents are also into handloom weaving. Theirs is a nuclear family and the family decisions are taken jointly in spite of secondary position of women Their annual income is about Rs.33,600/-.

She started a handloom weaving unit about one and a half month agro with a capital investment of Rs.50,000/-.

Figure 5.15: A view of handloom weaving unit.

She registered the unit before obtaining financial assistance from a bank. She employed 2 untrained workers.

Her family supports her in the endeavour. Her personal annual income amount to Rs.23, 600/-. Her achievement motivation is of less level. The locus of control is external. She has medium level stress levels associated with multiple roles. The levels of leadership qualities and the need for power are medium. While the task motivation is high, the risk taking behaviour is moderate.

5.22 Case Study 22

Mathammal is a resident of Thenpenmudi village in Karamadai taluk of Coimbatore district. She is 65. She is illiterate. She belongs to a backward class community. There are 6 members in her family. She is into handloom weaving. She used to work as a daily wage worker in the past. She has been in the present occupation for a year and a half. Her personal annual income amounts to Rs.23, 400/-.

She was born into a joint family. Her parents are illiterates too. They are also into handloom business. Their

Figure 5.16: A view of handloom weaving.

family annual income is about Rs.50, 400/-. Family decisions are taken jointly but women's position in the family secondary.

She is one of the 23 members of an Self Help Group. All of them were trained for 15 days by the government of Tamil Nadu. They registered their group on 26th November, 2006. Their capital investment is about Rs.20,000 and the financial assistance has been provided by Central Bank of India. The group manages all the business issues.

She receives all the support needed from her family. She has good social standing because of her entrepreneurship abilities.

5.23 Case Study 23

Sina Anjali is a 52 year old entrepreneur living in P.N. Palayam block of Coimbatore district. This middle school studied backward class woman used to be a daily wage earner before joining a Self Help Group which is into packaged drinking water business. Her personal income per annum is about Rs.16,000/-.

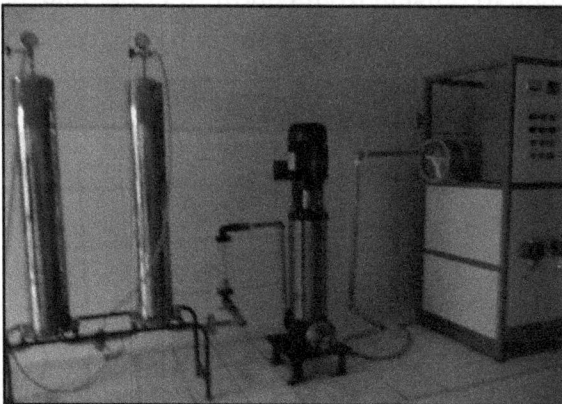

Figure 5.17: A view of packaged drinking water plant.

She is from a joint family. Her parents are also middle school educated. They run a canteen. Their income is about Rs.16,000/- per annum.

The Self Help Group, in which Sina Anjali is a member, consists of 4 members. Having received training for 6 days by an NGO, they started their own business with a capital investment of Rs.1,00,000/-. Their total assets value is about Rs.1, 25,000/-.

Anjali has been given much support by her family. Her social standing is positive. Her achievement motivation is medium and locus of control is internal. She has demonstrated medium levels of stress in playing multiple roles. Her leadership qualities as well as her need for power are of medium level. She has less level of task motivation and risk taking behaviour.

5.24 Case Study 24

B. Rahima, a 37 year old entrepreneur is from Jiyapuram village in Andhanallur block of Tiruchirapalli district. This backward class community woman changed her religion to Muslim from Hindu, after her marriage. She dropped out of school after studying 5th standard. She

Figure 5.18: A view of small hotel.

joined Prajasakhti Self Help Group in 2003. She is into a small hotel business.

Rahima's husband is a diabetic patient and is in need of a surgery. Her brothers also run small hotels and so Rahima has got some experience in managing such business. She employs 5 people in her business.

She was provided an amount of Rs.2, 50,000/- as loan from Indian Overseas Bank. She invested Rs.40,000/- on her own and thus the capital investment is Rs. 2, 90,000.

Her achievement motivation is medium and the locus of control is internal. The stress associated with the multiple roles of high level. Her leadership qualities and the need for power are high. She has medium level of motivation and her risk taking behaviour is risky.

5.25 Case Study 25

Sri Mani is a 32 year old entrepreneur from Nanjundapuram in P.N.Palayam block of Coimbatore district. She is a backward class woman and studied upto middle school level. There are 4 members in her family.

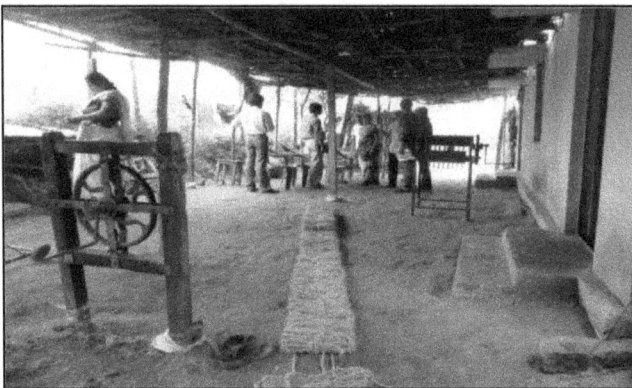

Figure 5.19: A view of rope making.

Before joining Sri Mani Amman Self Help Group as a member, she was a home maker. Her personal annual income is about Rs.16,000/-.

She was born in a joint family. Her parents are into rope making business. They also studied up to middle class level. Their annual income is about Rs.24,000/-. Women are active in the family and the family decisions are taken by them.

Central Bank of India extended financial assistance to Sri Mani Amman SHG. The group's capital investment is Rs.1, 50,000/-. The value of assets of the four member group is Rs.2,00,000. The members were trained for a week by an NGO. The members together carry out all the business related issues.

Sri Mani gets all the support from her family. Her family ties and priorities are good. She also looks after her children besides participating in the group's business. Her social standing is positive.

5.26 Case Study 26

G. Sharita (28) lives with her husband and only daughter in Kanuvai, Somayapalayam Road in P.N. Palayam block of Coimbatore district. The young high school drop out is from a backward class community. She was into a small home based business before joining as a member in Nandhavamam Self Help Group. Her annual personal income is about Rs.18,000/-.

She was born to middle school educated parents in a joint family. Her parents have a home-base cloth selling business. Their income is about Rs.18,000 per annum. Women take secondary position in the family because male members take the family decisions.

Figure 5.20: A view of home-based cloth shop.

She started a new venture close to her home. She registered the new firm in January 2006. Her capital investment is Rs, 1,00,000/-. She employed 4 untrained people. Financial assistance was provided to her by a private financial organization.

Her family supports her run the business. The growth rate of the business is considerably well. Her achievement motivation is medium and the locus of control is internal. Her stress level associated with multiple roles is less. Her leadership qualities and the need for power are of medium levels. Her task motivation is medium. She demonstrated less risk taking behaviour.

5.27 Case Study 27

S. Ambu is a 29 year old entrepreneur from Valliamankovil Strees, Vadaval Panchayat in P.N. Palayam block of Coimbatore district. She is a high school drop out. She is from a backward class community. She used to run a fancy store in the past but now she sells *vibhuthi* and *panchamritam*. Her personal annual income is about Rs.48,000/-. Theirs is a family of four.

Figure 5.21: A view of fancy store.

She was born in a nuclear family. Her parents had high school education. They run a fancy store at their home. Their income is about Rs.48,000/- per annum. Family decisions are taken jointly and the women's position is secondary.

She sells some of the other puja items keeping all her saleable items on a small bench in a street. She joined Sivanjani Self Help Group after some time. After joining the Self Help Group, she was given preference well as subsidy in the allotment of tenders. She was successful in the bid and she was allowed to sell the items on tender amount of Rs.1,00,000/- which she raised from personal resources and private finance firms. Her business has fluctuations but she does brisk business on festivals. She gets Rs.200 profit in a day in such items.

Ambu gets all the support from her family. Her social standing is positive.

5.28 Case Study 28

T. Rukmini is a 39 year old entrepreneur who lives in Jiyapuram village in Andhanallur block of Tiruchirapalli

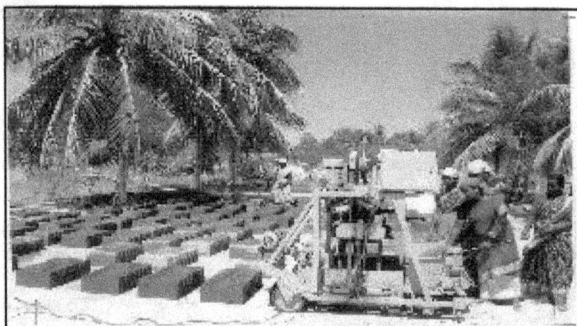

Figure 5.22: A view of brick manufacturing.

district. There are 4 members in her family. This backward class community woman studied upto high school level. She manages a hallow brick manufacturing unit set up by Maha Sakthi Self Help Group. Her annual income is about Rs.60,000/-.

She was nominated by the members of the group to manage the unit. The group has received an institutional credit to the tune of Rs.5,00,000/- from Indian Overseas Bank, Allur. The income from the business is shared among the members. She feels that if women have earning capacity they will never face problems in their lives.

She was opposed by her husband in early stages of her entrepreneurship but he is convinced after seeing her success.

5.29 Case Study 29

Savithri, a 55 year old middle school educated backward class entrepreneur is from Somayampalayam village in Periyanayakkam block of Coimbatore district. She lives with her husband and a child. She used to work as a daily wage labourer before setting up a small tea-stall which

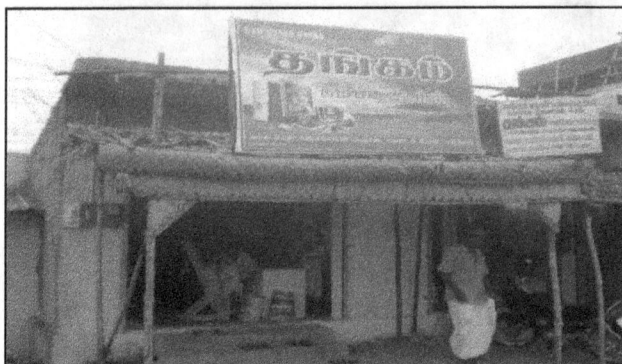

Figure 5.23: A view of tea stall.

she has been managing for about 10 years. Her personal annual income is about Rs.48,000/-.

She was born to illiterate parents in a nuclear family. Their income is about Rs.48,000/- per annum. Women take all important decisions and so their position is very active in the family.

Savithri and her husband worked on daily wages for about 15 years before establishing the tea-stall. They prepare breakfast along with tea in the morning. Their income and savings from their business was hardly sufficient for their children's education and weddings of their two daughters.

Having seen a tough time, Savithri joined Nandhavanam Self Help Group. Four members of the group are there now to run the stall, but the stall is not well furnished. They are looking for the financial assistance to furnish the stall and to extend the business for serving lunch orders.

The achievement motivation is less and the locus for the control is external. The stress associated with the multiple roles is medium. The leadership qualities are medium and they need manpower to expand the business. The task motivation is medium. Risk taking behaviour is less risky.

5.30 Case Study 30

Palaniammal, a 32 year old entrepreneur is living in Somayamalayam Panchayat, Periyanayakkam block of Coimbatore district. They are a family of ten. Palaniammal is from a backward caste and she studied up to middle school. She has 30 years of business experience. She owned a Prasadham stall near Maruthmalai Subrahmaniya Temple. Her personal income is Rs.24,000/-. She is also a member of SHG named as Sivanjali Self Help Group.

The group is operating 5 shops, where Palaniammal is incharge for one shop named *Arun Vilas Panchamirtham* Store. She is from a Joint Family. Her parents studied up to intermediate and they have a store of *panchamirtham* near the temple. Their earning is Rs.24,000 per annum. Decisions are taken jointly and the women's position in the family is secondary.

The enterprise is group activity and the nature is co-operative. They registered the enterprise on 2/07/2004. The

Figure 5.24: A view of Arun Vilas Panchamirtham

capital investment was Rs.2,00,000/-. They got the financial assistance form the private financial organization. The recognition is received at family level. The recruitment process is looked after by the whole group. The numbers of employees in the group are 14 and the values of assets are Rs.8,50,000/-. They received training for 6 months and the training was organized by the Panchayat union. They started with 2 shops and now diversified to 5.

Influence of the family members is considerable. Palaniammal's family members gave her full support in setting up the enterprise. They day to day operations are supported by the family members and group members. The enterprise is fully managed by the family members and also recruitment and management of labour fully managed by the group, they are having good growth rate. They received institutional finance as Rs.25,000/- from bank in that they also get Rs.10,000/- as subsidy. Coming to Family and social values, the family gives full support to her. The social standing of women entrepreneurship is positive. Savithri is playing multiple roles by taking care of her children also. The family ties and priorities are good. There are no instances in inhibiting the growth by the family members. Contribution of the family and society in her success are good.

5.31 Case Study 31

V. Kalamani, 45 year old entrepreneur living in Narasimhanaikem Palayam village of P.N.Palayam block in Coimbatore district. She is from a small family of 4 members. She is from a backward community and she studied up to high school. She has business experience of 2 years. Her occupation is Masala making business.

Kalamani's personal income is Rs.18,000/- per annum. She is from a nuclear family and her parents also studied upto middle school and they are also from the same business. Their earning is about Rs.60,000/- per annum. Decisions are taken jointly in the family and the women's position in the family is active.

She registered the enterprise on 14-12-2005 and the capital investment is Rs. 50,000/-. She got the financial assistance from the government. The recognition received about business is at community level. The recruitment process is looked after by the whole group. The number of employees in the group is 14 and the values of assets are Rs.50,000/-. The members received training for 6 months organized by the government. They started making pickles and it was diversified into making sweets in festival times and they now started selling cooked rice and cloth.

Influence of the family members is considerable. Kalamani's family members gave her full support in setting up to enterprise. Its day to day operations are supported by the family members and the group members. The enterprise is fully managed by the family. The recruitment

Figure 5.25: A view of enterprise.

and management of labour is fully managed by the whole group. Coming to family and social values, the family gives full support to her. The social standing of women entrepreneurship is positive. The family ties and priorities are good; there are no instances of inhibiting the growth by the family members. Contribution of the family in success is very good and the contribution of the society in their success is considerable.

5.32 Case Study 32

Padmavathi is a 39 year old entrepreneur. She lives with her husband in Poolavadi village in Gudimangalam block of Coimbatore district. She studied up to Intermediate. She is from a backward class community. She has been into handloom business for 4 years. Her annual income amounts to Rs.36,000/-.

She is from a nuclear family. Her high school educated parents earn Rs.36,000/- annually from their hotel business. Decisions in their family are taken jointly. Women's position in the family is secondary.

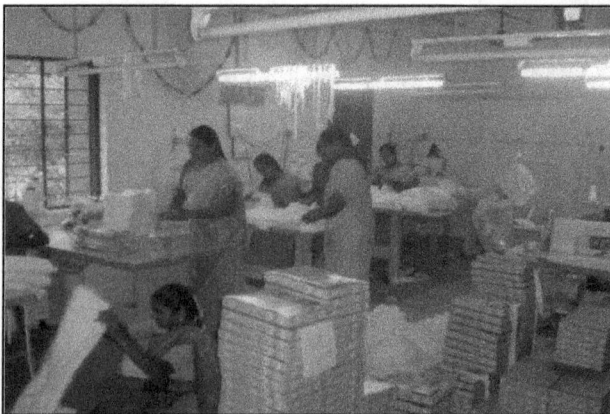

Figure 5.26: A view of handloom enterprise.

Her enterprise is a group activity and is of co-operative nature. It was registered on 18th February, 2003. The capital investment is Rs.40,000/-. There are four untrained employees. Financial assistance was provided by a bank.

Padmavathi receives much support from her family. She has her own ideas about empowering women. She empathises with women. She feels that co-ordinating women are not an easy task.

5.33 Case Study 33

Mylal, a 47 year old entrepreneur is a member of Jaihind Women's Self Help Group in Mogavanur village of Gudimangalam block in Coimbatore district. She is a schedule caste woman and has two married sons. She is illiterate. She makes brooms and broomsticks using thatch. Her annual income is about Rs.30,000/-. She lives with her sons and daughter-in-laws.

Her parents studied up to high school level. They also make brooms and broomsticks. Their income is about Rs.34,000/- per annum. Family decisions are taken jointly but women's position in the family is secondary. They have

Figure 5.27: A view of brooms and broomsticks making.

been into the business of making brooms and broomsticks for about 20 years. When the season is dull, they work on daily wages for living.

Mylal's business is a cooperative group activity. They registered their SHG in November, 2002. Their capital investment is Rs.95,000/-. They obtained loan from a bank. The unit was recognised as the 'Best Operation' and was given an award by the local Panchayat. There are 12 members in the group. All of them are trained.

5.34 Case Study 34

Jyothi, 36 year old entrepreneur, lives in Appallur village in Thottiyam block in Tiruchirapalli district. This middle school dropout woman is from a backward class community. She is into gem cutting business. Her annual income is about Rs.36,000/-.

She is from a nuclear family. Her parents are educated. They earn about Rs.60,000/- per year. Women's position in their family is active and the decisions are taken jointly.

The gem cutting business was started by a group with Rs.1,00,000/- as capital investment provided by a bank.

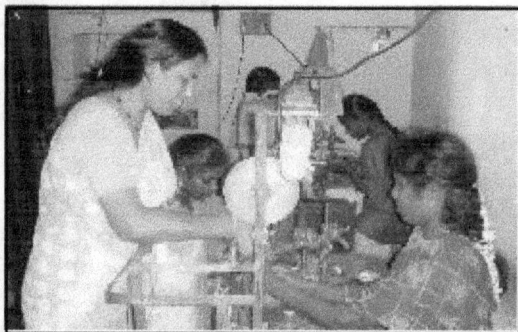

Figure 5.28: A view of gem cutting.

The growth rate of the enterprise is considerably good. All the 12 members of the group were trained in gem cutting for a month sponsored by an NGO.

5.35 Case Study 35

Amudha (36) is a founder member of Ajaimman Self Help Group. She lives with her family of 3 in Poolavadi in Gudimangalam block of Coimbatore district. This woman entrepreneur formed the SHG on 10th October, 2003. Her husband and she were into handloom business earlier but were not satisfied with limited profits. Her annual income is Rs.42,000/-.

She is from a nuclear family. Her parents had middle school education. They were also into handloom business. Their income is about Rs.42,000/-. Family decisions are taken jointly and women command respect in the family.

The Ajaiamman SHG has been efficiently managed by Amudha. The Self Help Group was provided with some financial assistance and some subsidy by the government. She feels that if women become economically independent, their families will lead quality lives. She gets much support

Figure 5.29: A view of handloom business.

from her family and neighbours. She received training on group management for 3 days. The training programme was conducted by an NGO.

5.36 Case Study 36

Lata (28) is an illiterate scheduled caste entrepreneur form Kosavampalayam in Udumalpet talu of Coimbatore district. Hers is a family of 4. She used to work on daily wages before venturing into selling *prasadham* and *vibhuthi* near a temple. She has been into this business for 5 years and 4 months. Her annual income is about Rs.25, 200/-.

Figure 5.30: A view of selling *Prasadham* and *Vibhuti*.

Her illiterate parents are daily wage earners. Their annual income is about Rs.61, 200. Family decisions are taken jointly and women's position in the family is secondary.

Lata's business is a group activity run on cooperative lines. Their capital investment is Rs.10,000/-. The met their investment from their own resources. The members of the group actively involved in the business proceedings.

Lata's family supports her very much. They family ties and priorities are good.

5.37 Case Study 37

D. Maheshwari, a 35 year old entrepreneur is from Thekkaampalayam village of Arumuthupalayam Panchayat in Coimbatore district. Hers is a family of four. She is from a backward class community. She studied upto middle school. She is into ready-made garment producing. Her annual income is Rs.30,000/-.

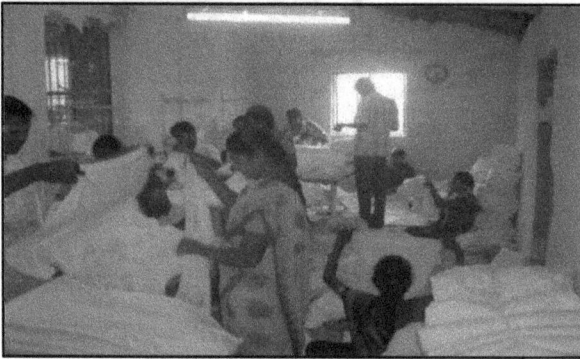

Figure 5.31: A view of producing readymade garment.

She is the founder member of Seventhi Self Help Group. The SHG was started 7 years ago, with the help of a local PDO. They registered the group's enterprise on 19th January, 2005 and the capital investment was Rs.3, 25,000/-. The financial assistance was provided by the Canara Bank. There are three untrained employees. The value of assets is Rs.1, 50,000/-. Maheshwari's family gives her all the support she needs.

5.38 Case Study 38

Valli Veezhili is a 40 year old entrepreneur who lives in Thottiyam block of Tiruchirapalli district. She belongs to a backward class community. She is a divorced woman. She has been into tailoring for about 10 years. Her annual income is about Rs.17, 500/-.

She is from a joint family. Her father is also a tailor. Women's position in their family is very active and they take all the decisions in the family.

Figure 5.32: A view of tailoring.

Valli, being a divorced woman, had to support herself to earn for living. Things improved after she joined a Self Help Group. She received 2 months training offered by an NGO. She started a tailoring business along with the other members of the group with a capital investment of Rs. 2,22,000/- provided by a bank as loan. The members also invested some money. The group members together make beds for infants.

5.39 Case Study 39

Manimegala, a 37 year old entrepreneur is living in Karadiviri village in Palladam block of Coimbatore district. She is from a backward class community. She studied up to middle school. Hers is a small family of four. Her annual income is Rs.12,000/- She used to be a daily wage earner.

Figure 5.33: A view of garment making.

Garment making is her preset activity. She is a member of an SHG.

The enterprise is a group activity and the nature of the activity is co-operative. The enterprise was registered on 27th January, 204. Their capital investment was Rs.5, 35,000 and the financial assistance was provided by the Canara Bank. There are 12 members in the group. The asset value of the unit amounts to Rs.5,00,000. The group members have not received any training.

The local PDO and the Sucha NGO are making every effort to help the group to be successful. They have a small place which is not convenient enough to work. The machines have to be run on single phase power. They have yet to reach their target levels. They are striving hard in spite of such problems.

5.40 Case Study 40

S. Gowri, a 36 year old entrepreneur is from Murugappa settiyar colony, Ganapathy Palayam village in Palladam taluk of Coimbatore district. She is from a backward class community and studied up to high school. She was a housewife before joining Soundeswari Self Help Group. The group started a garment industry with the help of Panchayat Board President and DRDA. There are four members in her family. Her annual income is Rs.24,000/-.

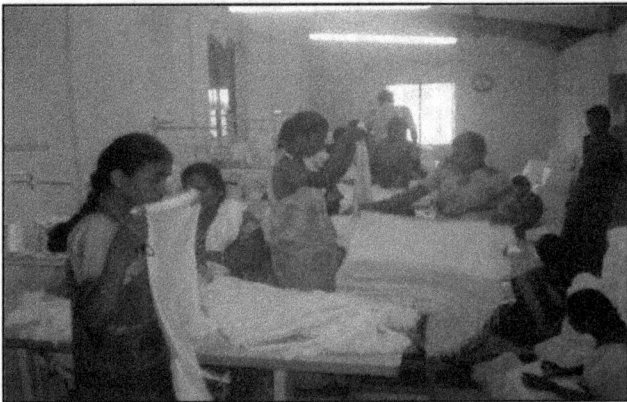

Figure 5.34: A view of garment industry.

The group registered the unit on 7th August, 2003 and their capital investment was Rs.3,00,000/-. The financial assistance up to the tune of Rs.2, 50,000/- was provided by a bank and Rs.50,000 was met from group members' savings. The group members were trained by DRDA. There are three employees. The value of assets amounts to Rs.3,00,000/-. Gowri's family supported her very much in setting up the enterprise.

5.41 Case Study 41

Kalavathi, a 45 year old woman entrepreneur, hails from Vakkampalayam Panchayat, Pollachi Taluka in Coimbatore district. She is into handloom weaving. She is from a backward class community and a middle school educated.

She joined "Sri Roja Self Help Group", which helped her getting financial help to grow her business. Hers is a family of three. Her husband helps her in her business. He helps her in purchasing raw material and in selling her products, sarees.

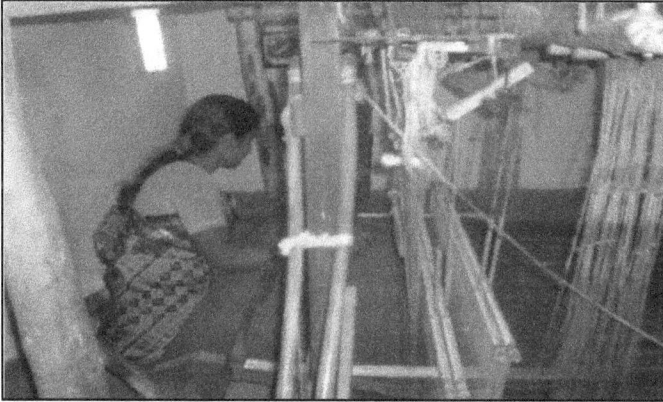

Figure 5.35: A view of handloom weaving.

She was born into a poor family and her parents were illiterate. They are daily wage labourers, who live in a joint family.

Having no business background, Kalavathi ventured into business with Rs.15,000/-, as capital investment. She has been trained in embroidery and doll-making. She is also being given some computer training organised by her Self Help Group with the help of the Panchayat.

She is playing the roles of a mother and a business woman. She has a medium level achievement motivation and her locus of control is internal. Her leadership qualities and task motivation, the both, are of middle level. She exhibited a moderate risk taking behaviour.

6

Emerging Issues and Challenges

Through the qualitative and quantitative analysis, the following psychological and sociological issues have emerged. They guide us through to understand the ground realities comprehensively. The study helped us to assess the challenges that would have to be faced and the strategies to be adopted to empower the women entrepreneurs in future.

6.1 Facilitating and Inhibiting Factors

The following Issues emerged from the sample districts.

☆ Women entrepreneurs want that there should not be any political interference in availing the government facilities

☆ Co-operation between the concerned members is a must

☆ The women entrepreneurs want tie-ups with the other organizations

☆ Government should increase the old age pensions

☆ More schemes should be introduced by the government to empower women

☆ Lot of SHG members want bank loans on 0 per cent interest

☆ Government should be more careful in selecting the beneficiary

☆ Entrepreneurs own investment is less and most of the investments are from banks or DRDA

☆ Government should put more focus on publicity of its schemes and facilities

☆ Schemes should be more transparent and corruption free

☆ Entrepreneurs expect more training to manage the company

☆ Many entrepreneurs have single phase electric line and they want three phase

6.2 Facilitating Factors

☆ Women entrepreneurs get support from the families and their husbands

☆ Women entrepreneurs get support from the local people

☆ Women entrepreneurs depends on a micro credit have low risk

☆ Good marketing facilities available for lot of SHG people

☆ Some of the group leaders train their own members

☆ It is a good move that lot of people get loans from the SHGs

☆ Women entrepreneurs get information from the NGOs about the government schemes

6.3 Inhibiting Factors

☆ Political influence should be reduced in sanctioning loans to Women entrepreneurs

☆ Women entrepreneurs do not receive proper response from the banks

☆ Some of the members financial status is low, so they are not able to run business

☆ Women entrepreneurs face problems like inordinate delays in getting bank loans

☆ Lot of backward caste faced problems people of the other community

☆ Lot of members face problems in getting shops through tenders

☆ In some of the enterprises like handloom weaving, members have to work for long hours causing back pain

☆ Some women entrepreneurs involved in food related business face problems from the health department

6.4 Constraints in Gender Development

In order to improve the position of women in India, it is necessary that the following impediments be addressed:

6.4.1 Illiteracy

Efforts to economically empower women have suffered due to their poor level of literacy. Approximately 60 per cent of the female population of India is illiterate, and hence, developmental efforts have to consider this aspect and particular efforts should be made to overcome this barrier. Identifying illiteracy as a key issue, the Government of India has launched a mass campaign to improve the literacy of children and adults. To ensure rapid progress and greater success in this literacy program, the participation of voluntary organizations would be used.

6.4.2 Ignorance

Many of the programs designed by the government have not been fully utilized by women because of the poor literacy. Ignorance of information on welfare programs, innovations in science and technology, etc. has become a major handicap. Alternate mechanisms have to be developed to reach a large section of this illiterate population in order to maintain productivity.

6.4.3 Conservative Nature

In many parts of India, women are very conservative and there is a need to develop ways to remove this cultural barrier through long-term approaches. It is necessary that developmental organizations consider this conservative nature of women, and ensure that this barrier is addressed by encouraging women to take part in various activities. Insecurity for women is a common phenomenon in many areas and there is a requirement to educate the community about the need for transformation and to increase women's mobility beyond homes. Here again long-term strategies are essential to ensure such a transformation.

6.4.4 Superstitions

Superstitions abound about both men and women, but the latter suffer more from these beliefs in many parts of the country. Welfare organizations interested in improving the social status of women have to take up this impediment seriously. Unless women overcome superstitions and social taboos; and adopt improved practices, poverty might remain as a perennial problem.

6.4.5 Poverty

Women are disadvantaged by the lack of education and knowledge brought about by poverty. Poverty leads to malnutrition, unhygienic conditions, and sickness in women. In Asian and African countries, ill informed and poverty stricken women have become an easy instrument for population explosion.

6.4.6 Organizational Support

Developmental issues in the fisheries sector of India have remained unorganized, and gender issues remain untouched. Developmental issues are often analyzed from the technical angle without due consideration of social questions. It is necessary that organizations view these social issues more intensely in order to solve the technical and social problems. Gender is one such major social issue which has to be given priority. Though there are several organizations taking care of men's welfare, there are few that look after the well-being of women.

These impediments are not confined only to the women of India, but also occur in several countries of the region, and hence, regionally coordinated efforts might help in speeding up the developmental process.

It is seen that the problems of women in entrepreneurship development stem from the three factors of the nature and quality of RESOURCES, CAPABILITIES and STRATEGIES. The prototype entrepreneur profile is of a man who is aggressive, extremely competitive, hard driving, and ingratiating, self-promoting, and high networking. The women counterpart is seen to be bestowing a naïve trust in good management, a reliance on 'natural' systems fairness, but suffers due to low benefit from networking, and low mentoring preparedness and skill. There is a perceived lack of preparedness to design and use resources that are rare, valuable, and hard to copy. Further problems are that women have less personal experience, knowledge, education training, and decision-making. They seem to have less personal integrity, less propensity for taking risk, and less need for achievement. Consequently, there is low feasibility and business desirability.

The paucity of role models and mentors is another significant factor in the entrepreneurship development of women. Personal factors like domestic commitments, child care and other care-giving responsibilities, constraints on personal liberty and on mobility, resistance from family (especially during launch stage) are key problems of women in general. But when dynamism and high mobility are indispensable factors for success in entrepreneurship, the effect is considerable for women in this sector.

Specific finance-environmental and organizational constraints for women in new ventures include low permanent and temporary working capital, low seed capital. Women also face greater difficulties than men in obtaining credits, in finding business partners, in accessing new markets and in getting information on business opportunities. Very few women are serial entrepreneurs. This gives the impression that they are not in a position to stake their new ventures with the rich rewards that might come from selling an enterprise. Many women choose entrepreneurship after having been out of the paid workforce for several years, after tending to the needs of a young family. A male-dominated labour market, institutions and policies, procedural delay in loans, apparent discrimination in selection for entrepreneurial development and training, unplanned growth of women entrepreneurship, the lack of specialized entrepreneurial programs for women, coupled with late commencement of women's entrepreneurship are other significant problems faced by women.

The general impression is that women do not invest sufficient capital in their own businesses: "Either they cannot or they will not." Women are perceived as lacking fundamental business skills and experience: no strong math skills and little or no relevant financial experience. Women are also considered bad business risks, as risk-averse and unable to take tough decisions. A new business venture, when launched, is seen as a response to *personal* life changes: either as a consequence of having children now in school, or as being recently divorced or widowed. Women entrepreneurs are thus considered 'reactive', not 'proactive'. Being out of the paid workforce for several years has negative consequences in the entrepreneurial world. Reduced cash reserves for investment are seen as a result of not having been serial entrepreneurs. Being out of touch with important business networks gives the impression of possessing rusty business skills. Lack of impetus from the lack of the 'positive pull' that men generally get from mentors, investors, customers, and the lack of the 'positive push' given by Education, Industry, Personal savings, Family assistance, current income, cash reserves, and mortgage able assets put women entrepreneurs at a distinct and serious disadvantage.

However, the fundamental problems as also the solutions lie in the educational, social and employment infrastructures rather than in problems with women's capabilities and commitment. Among the responsibilities of women as entrepreneurs, the imperative is rallying for gender-sensitive policies for SMEs. In particular there should be an awareness and readiness to facilitate articulation of problems, the actual presentation of problems to authorities, conduct of management programmes, entrepreneurial development programmes, encouragement of joint ventures among women. Moreover, it is their collective responsibility to demand job-oriented programs in women's Institutions, provide motivation to new entrants, encourage socio-cultural change, ensure easy access to credit and create conditions for gender-sensitive institutional infrastructure and gender sensitive policies. Special attention to rural women by reaching out through NGOs to arrange conferences and workshops, develop political awareness as well as to seek assistance from academic and research institutions to offer counseling and training are also imperative.

6.5 Recommendations

Though the above initiatives have been taken at the financial institution level, the issue of women's development is too big to be solved by these few steps. There is still a wide gap between the efforts and the actual needs at the field level. The financial support necessary to address this task is huge and, therefore, international agencies will have to support the national efforts. We feel that the following important steps should be given priority to help bring about the economic empowerment women:

☆ Provision of basic education for all.

☆ Liberal financial support to motivate women entrepreneurs.

☆ Incorporation of gender concerns in to every program.

☆ Evolution of long term marketing strategies.

☆ Selective and needs-based training for developing entrepreneurship and improving skills.

☆ Establishment of Fisheries Polytechnic Institutes to train women for specialized functions.

☆ Establishment of Self-Help Groups, Voluntary Agencies and Social Welfare Organizations and linking them to bank financing programs.

☆ Conduct of special workshops and seminars for better extension programs and education.

☆ Framing a policy for leasing smaller bodies of water to fisherwomen so that they can take up aquaculture as a commercial activity.

☆ Providing a common platform to all the developmental agencies, financial institutions, and research organizations to achieve an integrated approach for promoting women entrepreneurs.

☆ Increasing finance for programmes promoting production and entrepreneurship as sources of income.

☆ Creating of conditions allowing women to maintain sustainable sources of income and means of subsistence.

☆ Implement concrete social measures in support of households headed by women.

☆ Providing women with technical support, advice, further training and re-training in matters of transition to a market economy.

☆ Establishing community services where they do not exist or do not operate at full capacity in order to help rural women cope with housework.

6.6 Future Rural Entrepreneurs

The entrepreneurial orientation to rural development accepts entrepreneurship as the central force of economic growth and development, without it other factors of development will be wasted or frittered away. However, the acceptance of entrepreneurship as a central development force by itself will not lead to rural development and the advancement of rural enterprises. What is needed in addition is an environment enabling entrepreneurship in rural areas. The existence of such an environment largely depends on policies promoting rural entrepreneurship. The effectiveness of such policies in turn depends on a conceptual framework about entrepreneurship, *i.e.*, what it is and where it comes from.

The standard perception is that entrepreneurship is a special personal feature, either a person is, or is not an entrepreneur. According to this perception entrepreneurial traits, such as the need to achieve, risk taking propensity, self-esteem and internal locus of control, creativity and innovative behaviour, the need for independence, occupational primacy, fixation upon goals and dominance, are all inborn. Therefore, policies directed specifically towards promoting the development of entrepreneurship would not help much since such characteristics cannot be acquired much by training.

Another perception is that some cultures or some social groups are more conducive to entrepreneurial behaviour than others. According to this view, the factors that contribute to the supply of entrepreneurs are an inheritance of entrepreneurial tradition, family position, social status, educational background and the level of education. Based on research into the origins of business owners, it is believed that persons, who come from small business owner families, are more likely to become entrepreneurs than others. Studies of family position of existing entrepreneurs demonstrate that entrepreneurs are often found among elder children, since according to the explanation, they are pressed to take more authority and responsibility at earlier stages than younger members of the family. The outsider group, ethnic minority, or the outsider individual, the marginal person, who are by a combination of different factors rendered outsiders in relation to the social groups with whom they normally interact, are both viewed as a significant source of entrepreneurship. It is claimed that to minorities' small business ownership means escape form marginality. Whether educational background influences potential entrepreneurs or not is a matter of debate. The popular idea of an entrepreneur is that of a totally self-made man, lacking in formal qualifications. Apparently two things are involved simultaneously: propensity to start an entrepreneurial venture and skills to run the venture successfully.

A vital aspect of successful rural enterprise is preparing the next generation of citizens for the challenges and possibilities of starting businesses in their communities.

Although rural areas and individual entrepreneurs face unique hardships, they also have unique resources designed to encourage and support their growth. While rural communities may continue to lag behind in terms of national scores, employment rates and earnings continued entrepreneurial enthusiasm and the formation of more resources can only work to strengthen these regions.

6.7 Environment Conducive to Entrepreneurship

Behind each of the success stories of rural entrepreneurship there is usually some sort of institutional support. Besides individual or group entrepreneurial initiative the enabling environment supporting these initiatives is of utmost importance.

The creation of such an environment has already been started at the national level with the foundation policies for macro-economic stability and for well-defined property rights as well as international orientation. Protection of the domestic economy hinders instead of fostering entrepreneurship. National agricultural policies such as price subsidies to guarantee minimum farm incomes and the keeping of land in production during over-production periods are definitely counter-productive to entrepreneurship.

6.7.1 Financing Difficulties

Credit is available for women through a plethora of schemes but there are still bottlenecks and gaps. The multiplicity of schemes is not adequately listed nor is there networking among agencies. As a result, clients approaching one institution are not made aware of the best option for their requirements. A closely integrated data bank into which all agencies concerned are plugged is a real need.

6.7.2 Training in Skills

Appropriate training is still the key to a successful programme to develop entrepreneurship among women. There are funds available from several sources but finding effective trainers is the greater problem. NGOs like RUDSET in Karnataka have succeeded in achieving reasonably high success levels, but others including governmental bodies have still not reached these levels. Continuous monitoring and improvement of training programmes should eventually spread the cult of entrepreneurship among young women.

The findings from Gujarat clearly shows that training with respect to number of days, types of training skill development etc have been imparted, where as in Uttar Pradesh no such training has been given. No such interest has been shown by government or NGOs.

6.7.3 Confidence in Marketing

A major area of difficulty for women entrepreneurs is marketing. Several initiatives have been put in place to remedy this defect. At the initial stages women prefer to be locked into programmes which ensure almost total marketing support, since they seldom have the time or the confidence to seek out and develop markets. Even when they are otherwise in control of an enterprise, they often depend on male members of the family in this aspect. Marketing means mobility and confidence in dealing with the external world, both of which women have been discouraged from developing themselves by social conditioning.

A final area of concern in the case of women entrepreneurs is stagnation in their growth. This is due to various reasons like the demands of household duties, mobility problems and the need to expand space and staff. It is also often due to psychological causes like lack of self-confidence and fear of success (women who succeed often face hostility and resentment within their family circles). The necessary managerial and technical skills are also a barrier to the growth of women's businesses. Training and counseling on a large scale of existing successful women entrepreneurs who seem to have plateaued is a necessity.

6.7.4 Strategies Adopted for Enterprise Promotion

The policies and programmes targeted specifically to the development of entrepreneurship do not differ much with respect to location. From the perspective of the process of entrepreneurship, whether the location is urban, semi-rural or rural is not important in itself. For example, the needs of would be entrepreneur or an existing small business do not differ much from those in an urban area. To realize their entrepreneurial ideas or to grow and sustain in business, they all need access to capital, labour, markets and good management skills. What differs is the availability of markets for other inputs.

Involvement of the People's Organisations in all aspects of planning is need by the hour. Such organisations should work around the following.

☆ Giving preference to woman-headed families, women entrepreneurs, the landless, other poor families and traditional artisans

☆ Strengthening traditional artisans by providing them better tools and equipment, training and exposure. Assist them in establishing and strengthening backward and forward linkages

☆ Assessing the local market situation in terms of potential income from different forms of enterprise. This provides a practical basis for short listing the potential enterprises

☆ Facilitating the process of selection of enterprises by first-generation entrepreneurs and create awareness among them about feasibility factors, market demand and profitability

☆ Recommending selection of potential enterprises from those with a low gestation period and those involving minimal risk, because the participants belong to the poorest stratum

☆ Encouraging every member of the entrepreneur family to contribute their own might

The inputs into an entrepreneurial process, capital, management, technology, buildings, communications and transportation infrastructure, distribution channels and skilled labour, tend to be easier to find in urban areas. Professional advice is also hard to come by. Consequently, entrepreneurial behaviour, the ability to spot unconventional market opportunities, is most lacking in those rural areas where it is most needed *i.e.*, where the scarcity of 'these other inputs' is the highest.

These are the reasons why rural entrepreneurship is more likely to flourish in those rural areas where the two approaches to rural development, the 'bottom up' and the 'top down', complement each other. Developing entrepreneurs requires a much more complex approach to rural development than is many times the case in practice. It requires not only the development of local entrepreneurial capabilities but also a coherent regional/local strategy. Evidence shows that where this is the case, individual and social entrepreneurship play an important role in rural economic, social and community development. The top down approach gains effectiveness when it is tailored to the local environment it intends to support. The second prerequisite for its success is that ownership of the initiative remains in the hands of members of the local community. The regional development agencies that fit both criteria can contribute much to rural development through entrepreneurship.

6.8 Progress and Achievements

Most of the enterprises supported by the project are still nascent, their average duration by the mid-point in the project being about 1.5 years. However, they have already started bringing benefits to the entrepreneurs. Participants have earned, to differing extents, additional income and they have also experienced an enhancement in their social status. Where there has been an improvement in skills and equipment, the quality of services rendered by artisans has improved and it increased self-confidence among the entrepreneurs. With some productive physical asset in their hand and empowered with a livelihood skill, the poor families have become optimistic about their future. The more enterprising among them are planning to increase other activities because they now have greater risk-bearing capacity.

6.9 Challenges and the Future

Two major areas that need strengthening vis-à-vis newly introduced enterprises are literacy and training in entrepreneurship development. Despite the present good performance, much has to be done in the area of record-keeping. This weakness is mainly due to poor literacy among the women entrepreneurs. Most of the entrepreneurs are landless or marginal land-holders and belong to the poor stratum of society. Obviously, they also constitute the lowest stratum in respect of literacy. It was observed that the entrepreneurs need systematic training in entrepreneurship development. Such training will surely be useful for all entrepreneurs to run their businesses more profitably and efficiently. This will enable them to reach higher orbits of business. This will be especially useful for more creative and ambitious entrepreneurs. This will also enable the participants to undertake marketing and withstand competition.

The success of the land-based and livestock-based interventions is expected to increase the purchasing power of the participant families. This will increase expected demand for and profitability of new enterprises in the area. They would also need additional funds to upscale their business as individuals on a credit basis could meet this need of funds.

Gender inequality in sharing economic power is another factor accentuating female poverty. Prevailing micro-economic policies require rethinking as they emphasized the formal economy to the virtual exclusion of other sectors, restrained initiatives taken by women, and did not distinguish between problems of men and women. Gender analysis of political platforms and socio-economic programmes is an important tool to formulate a poverty alleviation strategy. Such a strategy should aim at better protection of women's health, sound nutrition in early childhood, early development of the child, equal opportunities in employment and promotion, better care for the pregnant, old-age, and sick women, improved rural education and social protection, and wider participation of women in decision-making. The home economy reflects the century-old division of work between the sexes. Working women come home from the field, animal farm, plant, school, or hospital and start their "second shift" while men hardly ever participate in the education of children or work around the home. The traditional work division in the household impacts significantly on the self-sufficiency of women, gender equality in the labour market, efficiency of labour policies, population trends, and the socio-economic situation of the country in general.

As society gets more structured and solutions are found for the demographic, environmental, social, economic, political, attitudinal and household problems, educational levels will inevitably grow. Higher education standards of the nation and society would lead to the use of technologically more advanced methods. Knowledge held by people is part of their personal and social capital. The wealth of the developed countries is based on a highly educated population. Education and knowledge are essential for the creation of material and cultural goods. They are pre-requisites for a sustainable, stable and democratic society. Education allows a human being to feel free, independent and responsible.

Access of women and girls to secondary and higher education is necessary if more women are to become socially active. The strengthening of their socio-economic status largely depends on their education and the standards of general and professional training. Literacy and education are the basic means of improving their health, nutrition, education and family status; defending their rights and freedoms; expanding their inclusion in social and political life, and securing equal access to decision-making. Education produces knowledge and helps understand the world outside the household and family. It helps achieve self-sufficiency and economic independence, and produces a feeling of self-confidence and dignity. Low educational levels are generally accompanied by high rates of childbirth stemming from ignorance on family planning. Dire consequences may occur in the absence of conscious efforts to support today's girls who will tomorrow become women, mothers, workers and wives. Their educational, professional and cultural levels will alone shape the social environment.

6.10 Conclusions

Women represent a significant resource of any country. The transition to a market-based democratic society has undermined the social status of the majority of the population. Women, children and families suffered the biggest damage. Women were never able to draw the full benefits of socio-economic and political opportunities existing in this country, much less in an oriental society imbued with the idea of inherent inequality of sexes.

The transition to a market economy makes things worse. Saddled with numerous social responsibilities and constrained by lack of equality, women easily lose whatever advantages they had. The condition of rural women deserves special attention as many of them face specific problems related to the rural way of life, economic hardship, agricultural technology and local customs.

Socio-economic problems are especially affecting women who are the sole breadwinners in the family. They and their children are the likeliest victims of the widespread impoverishment. Poor education and training force them to agree to a mere pittance.

The economic transition has directly affected women's status in society. Enormous problems have arisen in the employment of rural women. In the past, women's status in society and economy depended on rules defined by the state. Today the influence of the state is waning. Social allocations will inevitably shrink and state interventions in any area will be on a much smaller scale. Women should be more active in defending their interests. They should abandon unrealistic and unreasonable expectations. Women should also be more active in their search for solutions to social, economic and political problems.

References

Anonymous (2006). Socio – Economic Review, Gujarat State 2005 – 06.

Anonymous (2007). *Gujarat: destination of Choice.* Vibrant Gujarat Global Investors' Summit 12 – 13 January.

Chowdhary and Prakash (1997). *Entrepreneurial Motivation-Factors and Features.* Ajmer, Center for Entrepreneurship and Small Business Management, Maharshi Dayanand Saraswati University.

Devi and Thangamutha (1997). *A Case Study of Women Entrepreneurs in Tiruchirapalli District.* Trichirappalli, Bharathidasan University.

Ganesan,S.(2003). *Status of Women Entrepreneurs in India.* New Delhi, Kanishka Publications.

Ganesan, R (2005). *Psychosocial Disposition towards*

Emancipation of Women Entreprenuership - A Study on Women Entrepreneurs in Food Processing Enterprise., 1st International Biennial Conference, Centre for Entrepreneurship and Small Business Management, Maharishi Dayanand Saraswathi University, Ajmer, Rajasthan, Sep 9-11, 2005.

Histrich, D. Robert and Brush, G. C. (1986). *The Women Entrepreneur*. Lexington, Massachusetts, Lexington Books.

Husain, Akbar and Khan Mohd.Ilyas. (2001). Motivational factors for Rural Development Personnel. *Journal of Community Guidance and Research*, 18(1), 17-24.

Jones, L. and I. Sakong. (1980). Government, Business and Entrepreneurship in Economic Development: Korean Case, Cambridge, MA: Harvard University Press.

Kaza P. Geetha. (1996). *Women Entrepreneurs and Bank Credit – Problems and Perspectives*. Programme on Gender issue in Credit disseminations (26-29 August, 1996) BIRD, 1996 (NABARD).

Kumari, Souda. (2000). Women's empowerment. *The Educational Review*, 106 (11), 190-191.

Misra, Rajeshwar. (1995). Personal characteristics and strategies as factors of rural leadership effectiveness. *Journal of Rural Development*, (Jan.), 11(1), 59-68.

Muthayya, B.C. and Loganathan, M. (1990). Psycho-social factors influencing Entrepreneurship in rural areas. *Journal of Rural Development*, Vo.l 9(1), pp. 237-281.

Maps of www.mapsofindia.com

Preetha and Parthsarathy. (1999). *Entrepreneurship Development of Women at Micro Level*, Centre for Adult,

Continuing Education and Extension Institute for Entrepreneurship and Career Development

Pujar V.N. (1989). *Development of Women Entrepreneurs in India: Entrepreneurship Development in India*. Sami Uddin Mittal Publications.

Rani, D. Lalitha. (1996). *Women Entrepreneurs*, APH Publishing House.

Singh Satvir, Personality. (1997). Motives and Entrepreneurial Success, *IJIR*, Vol. 33 No. 2 Oct.

Stevenson, H.H, *et al.* (1985). *New Business Ventures and the Entrepreneur*. Homewood, Irwin.

Thakur. (2000). *Explaining Entrepreneurial Success: A Conceptual Model*. Kolkata, Indian Institute of Management Calcutta.

Timmons, J.A. (1989). *The Entrepreneurial Mind*. Andover: Brick House.

Tiwari, J. (1998). Indian Social Order of Emerging Values. *Indian Journal of Psychometry and Education*, 29(1), 13-16.

Tyson, L, T. Petrin and H. Rogers. (1994). Promoting Entrepreneurship in Central and Eastern Europe. *Small Business Economics* 6, pp. 1- 20.

www.ibef.org